The Timing of Biological Clocks

THE TIMING OF BIOLOGICAL CLOCKS

Arthur T. Winfree

An imprint of Scientific American Books, Inc.
New York

This book is number 19 of a series.

Library of Congress Cataloging-in-Publication Data

Winfree, Arthur T.
The timing of biological clocks

(Scientific American library; 19)
Bibliography: p.
Includes index.
1. Circadian rhythms. 2. Biological rhythms.

I. Title. II. Series.
QH527.W54 1986 574.1′882 86-15602
ISBN 0-7167-5018-X

Printed in the United States of America

Book design by Malcolm Grear Designers

Scientific American Library
An imprint of Scientific American Books, Inc.
New York

Distributed by W. H. Freeman and Company,
41 Madison Avenue, New York, New York 10010
and 20 Beaumont Street, Oxford OX1 2NQ, England

2 3 4 5 6 7 8 9 0 KP 5 4 3 2 1 0 8 9 8 7

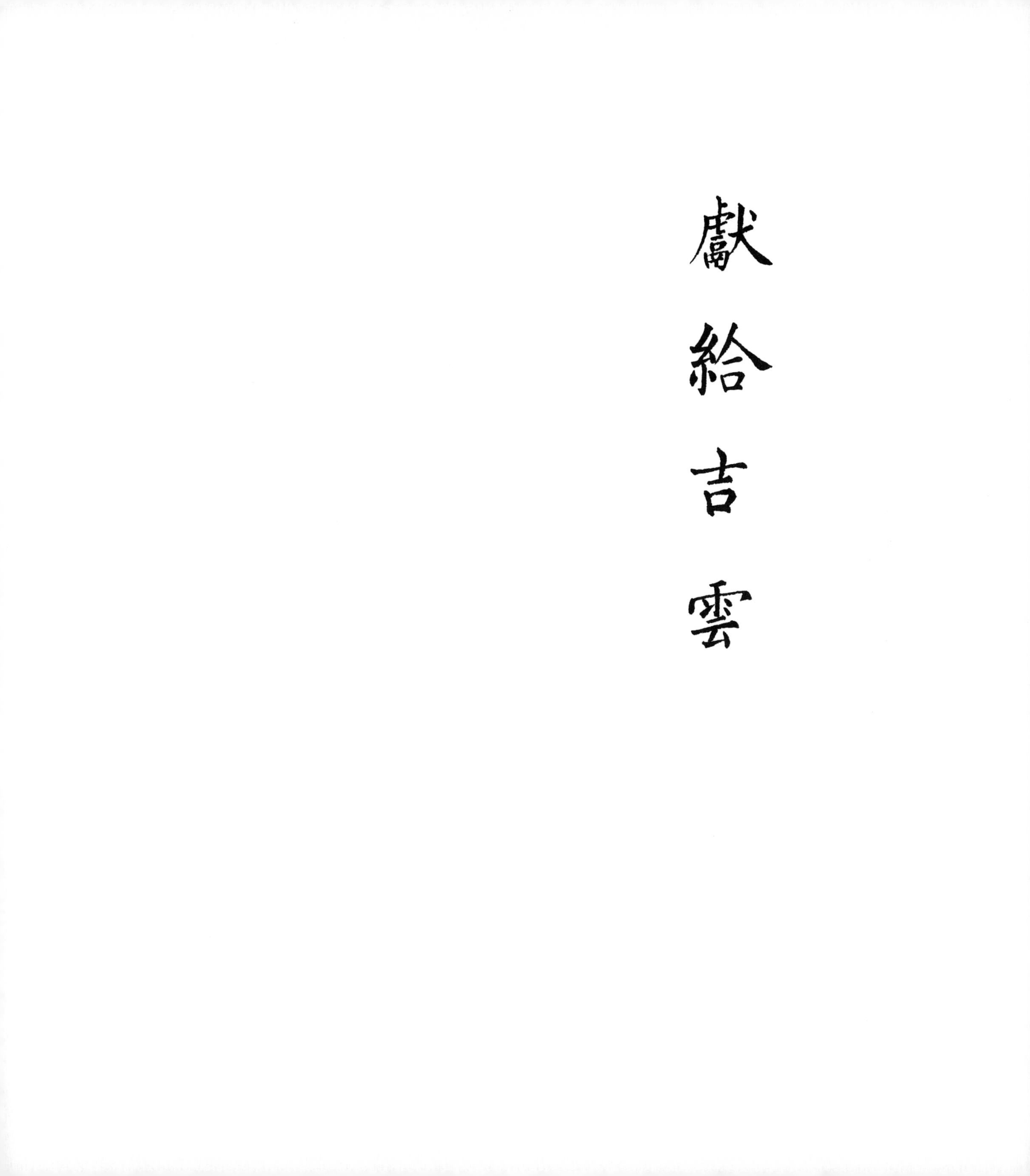

獻給吉雲

Contents

Preface

In a sense everything happens by accident. Any deliberate event is guided by a purpose that selects among accidents, ignoring some and turning others into opportunities, but the outcome is always colored by the particulars of those accidents.

This book can be traced back to a certain misdirected letter of fall 1963. I was then studying engineering physics at Cornell University as preparation for a research career in biology and needed a summer job for lab experience. Having decided to apply to the Marine Biological Laboratories at Woods Hole, I asked my engineering program director for someone there to contact. Trevor Cuykendall of course had friends at Woods Hole, but to him that meant the Oceanographic Institution. The possibility of confusion occurred to neither of us, so my eager entreaty went off to "Dr. J. B. Hersey, Marine Biological Laboratories, Woods Hole, Massachusetts." Nevertheless it found him, in the midst of outfitting the RV *Chain,* a geophysical research vessel, for the International Indian Ocean Expedition. He needed crew. In exchange for standing watches and looking after sonar electronics I could play Charles Darwin on the H.M.S. *Beagle,* as ship's biologist by default!

Sonar pings echoed endlessly through the days and nights of summer at sea. They often returned, purified and resonant, from mysterious strata that descended by day and rose by night, as our ship skimmed the interface between waves and wind. Jacques Cousteau chased those deep clouds in his diving saucer and encountered only occasional small fish, whose swim bladders, it was guessed, might resonantly oscillate when pinged by an acoustic impulse. But why the vertical migration of these deep scattering layers? A biology book in the ship's library told of a widespread behavior recently christened "circadian rhythmicity." Was the layer of fish rising and falling in response to an internal cycle? The eventual answer was probably not: they just avoid light in the daytime by descending, perhaps in pursuit of smaller prey who prefer darkness. But as I tried to solve that riddle, after weeks of concentrating on time zones, polar-coordinate navigation, sonar impulses, and acoustic oscillators, an eclipse started an avalanche of free association. My thoughts were led immediately to populations of oscillators, mutual synchronization, and resetting of amplitude and phase by impulses. The result (five years later) was a 400-page document intended as a Ph.D. dissertation in experimental biology. Though it was rejected, many chapters were rewritten after another five years of experiments with yeast cells and fruit flies and yet another five years with fungi and chemical clocks, to appear as *The Geometry of Biological Time* (Springer-Verlag, 1980).

The Scientific American Library was just getting started then. A senior editor there, Peter Renz, suggested that the technical monograph of 1980 might be turned into something more readable, and I jumped at the opportunity to be associated with the well-known scientists who were already writing for the series. The book was outlined during a week of backpacking the Na Pali coast of Kauai in the summer of 1981. Another five years were about to begin, during which many changes took place, not the least of them being the book's directions and contents. Its original more technical second half, extending the theme of geographical time zones to interpret spatial patterns of timing, found peace at Princeton University Press in November 1985 *(When Time Breaks Down),* while work continued on making the first half untechnically lucid for the Library. Any success in that direction should be credited to Andrew Kudlacik, my patient editor during the final year, and project editor Susan Moran.

The process of evolving theory, at least as I practice it, seems to consist mostly of accidents and misconceptions, but persistent refinement through laboratory experiments and through editing by the Scientific American Library team has winnowed out a beautiful pattern. I owe thanks, as always, to the National Science Foundation, for patronage almost ever since RV *Chain* tied up again in Woods Hole; to Purdue University for sabbatical leave; and to the generosity of many hosts during the book's five-year gestation while I sought refuge from midwestern allergens. This emulation of a pinball machine began in winter 1981 at J. T. Enright's desk in Scripps Institution of Oceanography overlooking the Pacific, continued with George Oster in the entomology department at U.C. Berkeley and amid cherry blossoms in Osaka, where I was the guest of Ryoji Suzuki and the Japanese Society for the Promotion of Science. The next springtime found me with J. D. Murray at his Centre for Mathematical Biology in Oxford, and the spring afterward, during a John Simon Guggenheim Fellowship, at my parents' cottage on Longboat Key and at Los Alamos National Laboratory as guest of George Bell. Final editing was completed overlooking Scripps Pier again, in La Jolla, during a John D. and Catherine T. MacArthur Fellowship in 1985–86 as the guest of Henry Abarbanel and the Institute for Nonlinear Science at U.C. San Diego. The book is finished, and so is the bouncing, thanks to a welcome into the Faculty of Ecology and Evolutionary Biology at the University of Arizona, where my laboratory will resume its winnowing of accidents.

ARTHUR T. WINFREE
Tucson, Arizona
Summer 1986

The Timing of Biological Clocks

In the year 1525, his health recovered three years after return to Europe, Antonio Pigafetta translated his ship's log into French, illustrated by his own painting.

Chapter One

Time Zones

. . . it was Thursday,
which was a great cause of wondering to us, since with us
it was only Wednesday.

ANTONIO PIGAFETTA
9 JULY 1522

A hardy little ship rides at anchor off the Canary Islands in midsummer of 1522. Fernão de Magalhães' (Magellan's) westward expedition to circumnavigate the earth has limped home, its captain buried long ago in a world previously unknown to Europeans, its original crew of hundreds reduced by hardship and disaster to 31 ailing men, all aged beyond their years. As he pens almost the last of three years' daily entries, a warm offshore breeze ruffles the pages of Antonio Pigafetta's logbook:

> In order to see whether we had kept an exact account of the days, we charged those who went ashore to ask what day of the week it was, and they were told by the Portuguese inhabitants of the island that it was Thursday, which was a great cause of wondering to us, since with us it was only Wednesday. We could not persuade ourselves that we were mistaken; and I was more surprised than the others, since having always been in good health, I had every day, without intermission, written down the day that was current.

Thus did sixteenth-century Europeans first encounter a phenomenon that excited them with the wonder we feel today about the intertwining of space and time in relativity. They eventually came to accept that this quirk of time was a consequence of traversing space, just as they had to accept that you can sail west and return from the east: everybody knew it in theory, but few were quite prepared to find that it had become a commercial reality.

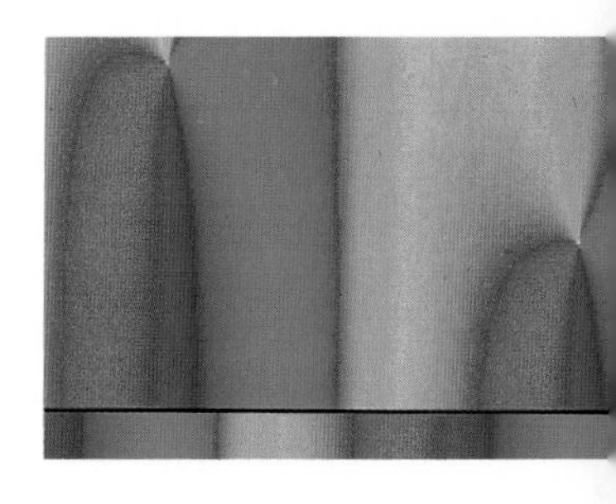

As global commerce expanded, the paradox of the vanishing day became more and more practically annoying. More than three centuries after Pigafetta's return, the young Charles Dodgson (the future Lewis Carroll, author of *Alice in Wonderland*) suggested what to do about the paradox. A clergyman's son, he amused himself, family, and friends by writing a humorous periodical called *The Rectory Umbrella.* About 1850 it carried the following whimsical summary of Pigafetta's Paradox:[1]

> Half of the world, or nearly so, is always in the light of the sun: as the world turns round, this hemisphere of light shifts round too, and passes over each part of it in succession.
>
> Supposing on Tuesday it is morning at London; in another hour it would be Tuesday morning at the west of England; if the whole world were land we might go on tracing Tuesday morning, Tuesday morning all the way round, till in 24 hours we get to London again. But we know that at London 24 hours after Tuesday morning it is Wednesday morning. Where then, in its passage round the earth, does the day change its name? Where does it lose its identity?
>
> Practically there is no difficulty in it, because a great part of its journey is over water and what it does out at sea no one can tell: and besides there are so many different languages that it would be hopeless to attempt to trace the name of any one day all round. But is the case inconceivable that the same land and the same language should continue all round the world? I cannot see that it is: in that case either there would be no distinction at all between each successive day, and so week, month etc. so that we should have to say "the Battle of Waterloo happened to-day, about two million hours ago," or some line would have to be fixed, where the change should take place, so that the inhabitant of one house would wake and say "heigh ho! Tuesday morning!" and the inhabitant of the next, (over the line,) a few miles to the west would wake a few minutes afterwards and say "heigh ho! Wednesday morning!" What hopeless confusion the people who happened to live on the line would always be in, it is not for me to say. There would be a quarrel every morning as to what the name of the day should be. I can imagine no third case, unless everybody was allowed to choose for themselves, which state of things would be rather worse than either of the other two.

Thus today we have the international date line, relegated like an embarrassing error correction in 1884 to a meridian as far as possible from Greenwich, England. We also have astronauts circling the globe in 90

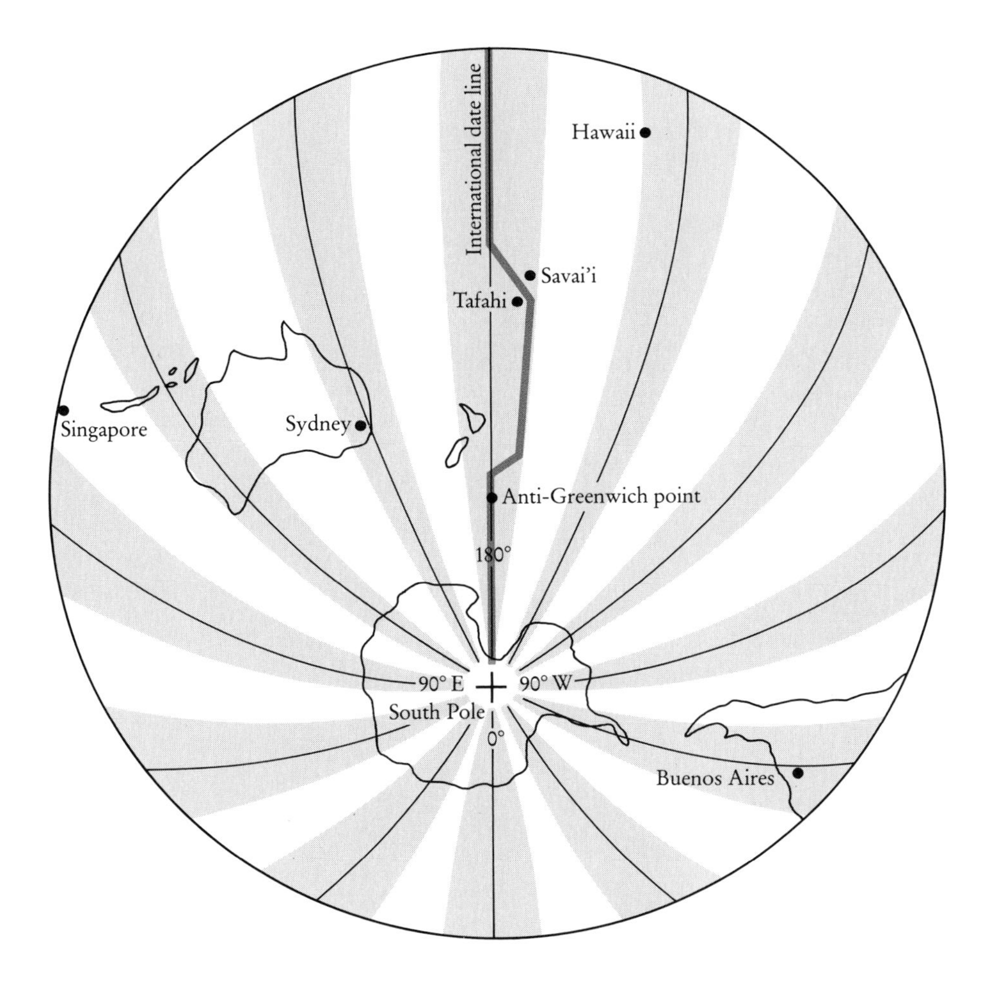

Citizens of Tafahi and Savai'i deal every day with the paradox that citizens of Greenwich pushed as far as possible from their own immediate concerns.

minutes, whose travels show us Pigafetta's paradox in its most extreme form. If an astronaut were to reset his wristwatch every 4 minutes as he crosses another time zone from west to east, he would gain nearly a full day in the 90 minutes it takes to get from Hawaii back to Hawaii. But when he crosses that line in the middle of the Pacific, he nominally pops back 24 hours into "yesterday," so he arrives back over Honolulu 90 minutes later *today,* not tomorrow. Each new date starts along the date line as it rolls through midnight. The international date line leads an expanding crescent of tomorrow around the east side of the earth into the dawn light, around through noon and dusk, and back to midnight. When the date line reaches the midnight point it starts the next day. So

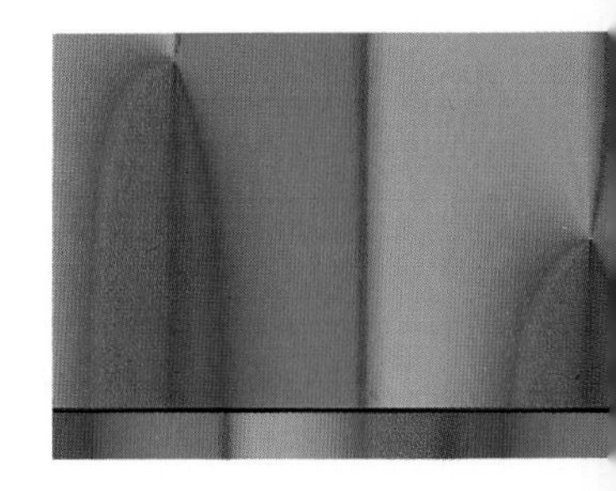

the citizens of Tafahi, in the Tonga Islands, are among the first to report back to work after each weekend. Meanwhile, their neighbors on Savai'i, in nearby Western Samoa, 150 miles to the east, greet the same moment as a Sunday morning, exactly as Lewis Carroll foresaw.

By this device, Europeans accommodated intellectually to the fact that the earth is a rotating ball. Merely intellectual accommodation was sufficient for that era, but later the fact also turned out to have physiological consequences that could not be dealt with so easily. Pigafetta was not their discoverer. At the end of his trip he remained exactly one day behind the rest of Europe, there being no way known at the time for him to catch up other than circumnavigating the globe again in the opposite direction. This full day's lag may have bewildered Pigafetta's mind but it did not affect his body: a gradually acquired shift in time by exactly one day has no known consequences for our physiological routines. Pigafetta functioned normally on local time even if his log showed the wrong day.

No individual noticeably felt the effects of shifting time zones until about 400 years later, when in 1931 Wiley Post flew eastward around the world in eight days. It turns out that your body cannot adjust to changing time zones much faster than two hours a day, as though your

Wiley Post used the Winnie Mae in the early 1930s to explore the effects on pilot performance caused by resetting his circadian clock.

skin can travel at arbitrary speed but your insides are limited to about 100 miles an hour. By flying across time zones much faster than that, Post discovered that he had an internal clock; he recognized the adverse effects of time zone displacement on his flying proficiency, and he struggled to evade them. He was the first man to experience jet lag, that disconcerting sensation of time travelers that their organs are strewn across a dozen time zones while their empty skins still forge boldly into the future, be that tomorrow or yesterday on the local calendar.

Post had a clock, or we might say, *was* a clock because he evolved on this planet. We live on a rotating planet. We grew up here, straight out of the warm Archeozoic seas where molecules first assembled into genes, and genes into species. For three billion years, life here has grown and adapted, passing from cell to cell innumerable times in unbroken descent, generation after generation. We and all other living things are the aggregate of all the changes made in that descent. All the while, we've felt the sky brighten and darken again and again while the planet relentlessly rotated: a trillion cycles of brightness and dark, of warmth and chill, never missing a beat, always felt deep in the chemical essence of what we are. We are well adapted to the pervasive monotony of sunrise and sunset, to the steady tone of a planet tirelessly spinning.

What would a trillion cycles sound like? Like high C for 400 years. Little wonder, then, that we've grown used to it, that we harmonize deeply with that unending note. Little wonder, too, if, as we first set foot beyond our home and venture out among the stars, we still hum with the pitch of our homeland. Anyone we meet out there will know where we came from, not just by our carbon and our water and the colors we see best, but also by the approximately 24-hour pitch of all that we do.

Keeping Time with the Earth

Physiological time, like local time on a rotating world, also has a circular character. Resetting any clock, internal or external, by one or several full cycles has no observable effect. But resetting a biological clock within a cycle does have physiological consequences, as the phenomenon of jet lag shows. Shifts within the cycle are called shifts in *phase*, the position of a repetitive process in its cycle (for example, a phase of the moon).

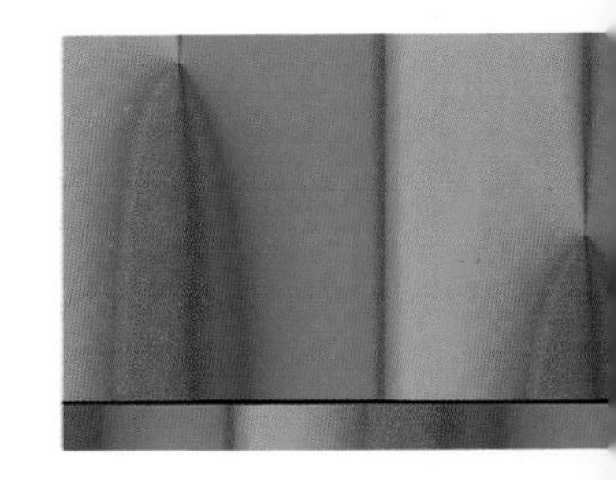

Quite apart from the new experience of jet travel, some means to reset the phase of a biological clock is made necessary by the slight mismatch

BIORHYTHM

There is nothing so absurd but some philosopher has said it.

CICERO, *DE DIVINATIONE*

The fact that living things on this rotating planet harbor internal clocks is easy to believe. This is surely one reason that the pseudoscientific Biorhythm theory has gained widespread acceptance, even among well-educated people. Its basis is the claim that we all have cycles comparable to a woman's menstrual cycle, but much more precise: physical ability waxes and wanes every 23 days, emotional condition varies in a 28-day cycle, and intellectual performance may follow a weaker 33-day cycle.

The three periods are the same for everyone, regardless of age, sex, or medical condition. Each period is an exact integer number of days, as though some inner mechanism were counting day/night cycles. Otherwise the theory would have to be dismissed *a priori:* the delicate timing of two- and threefold coincidences would be drastically undermined over the course of a long life if the three periods were numbers such as 28.02 ± 0.04 days (depending on the individual, vicissitudes of health, travel, and so on) rather than 28 discrete clicks of the earth-sun clock. Anyway, the putative cycles start on the day of birth, whether chosen by the Caesarean surgeon or by the baby and mother in the usual way, whether or not the process spanned a midnight. Presumably the effective day of birth for Biorhythmic purposes must be offset by the cumulative number of date line crossings in a lifetime. Individuals who have missed a lot of days while exploring caves, living above the Arctic Circle, or working in submarines would presumably also need to make an appropriate correction to their actual birth dates. It is not clear what should be done if you move to a substantially different time zone.

A day when two or three of these rhythms are passing their average levels (heading up or down) is said to require particular care: you are more accident prone. Days when two cycles are both negative or when one is cresting while the other is at trough are also alleged to be dangerous. The pattern of such days, starting from birth, is exactly the same for everyone. The dominant physical/emotional (male/female) part repeats exactly after 644 days. Including the 33-day intellectual cycle, the whole sequence of critical days repeats from a second effective birthday at age 58.2 years.

Biorhythm was heavily advertised and became a commercial success long before anyone checked to see whether it works. The notions behind this "science" originated just before the turn of the century in the mind of Wilhelm Fliess, a nose and throat surgeon in Berlin, author of a monograph lengthily entitled in German "Relations Between the Nose and the Female Sex Organs . . . " The cause was taken up shortly afterwards (precipitating rancorous priority disputes) by Hermann Swoboda, a psychologist at the University of Vienna. Fliess and Swoboda developed the numerology of the 23-day "male" cycle of physical ability and the 28-day "female" cycle of emotional condition (not the same as the more real but less exact menstrual cycle). The less important 33-day "intellectual" cycle was invented in the 1920s by Alfred Teltscher, an engineer in Innsbruck. There was a renaissance of popular interest in the United States during the 1930s, but it didn't survive World War II. Few living persons have read the authoritative source, Teltscher's *The Rhythms of Life: Foundations of an Exact Biology,* published in 1906. Martin Gardner, author of *Fads and Fallacies in the Name of Science,* reviewed Fliess's works for *Scientific American*'s "Mathematical Games" section of July 1966, and dismissed them as arbitrary nonsense, dusty "masterpieces of Teutonic crackpottery." However, George S. Thommen took a more constructive approach by reviving this lore in the United States. His books *Is This Your Day?* (1964) and *Biorhythm: Is This Your Day?* (1969) have gone through numerous revisions and reprintings, and Thommen became president of a firm that sells charting kits and calculators to the gullible. Everyone has by now seen advertisements for "computerized" Biorhythm charting services. Casio Biorhythm calculators are commonly seen in the hands of well-dressed individuals in airports and office buildings.

The notion clearly has appeal, and it has earnest supporters. Unfortunately, their evidence, largely anecdotal or based on inadequately controlled statistical surveys and unpublished sources, would not stand scrutiny. Nonetheless, a fair-minded person could persist in asking whether the Biorhythm theory might be true anyway; important discoveries are sometimes made intuitively or by diffuse folk wisdom before formal proof can be mustered. The airlines, the military, and others responsible for industrial accident prevention thought that a rigorous check might be worth the trouble. During the 1970s in a dozen independent published studies, close to 40,000 care-

(continued on next page)

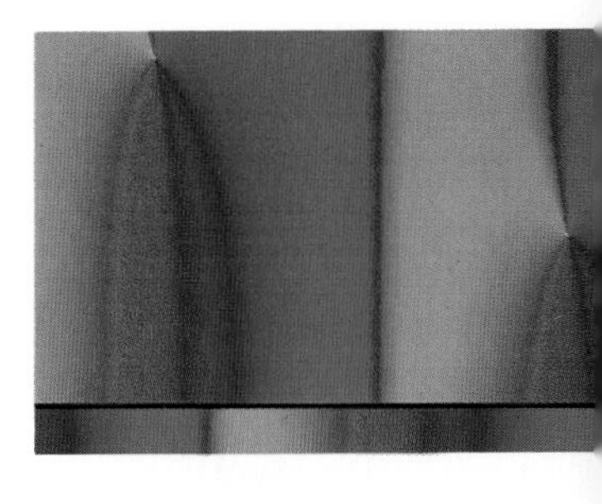

fully documented suicides and accidents of diverse kinds were compared with the predictions of Biorhythm, according to birthdays. No correlation was found.[2] As one team concluded, "Individuals do have good and bad days; their occurrence, however, is not predictable by the Biorhythm theory." People experience or report more trouble on the predicted days only if they have been primed to expect it on those days.

---

between that clock's native period and the period of the earth's rotation. A discrepancy of an hour or so one way or the other is usual among the many species whose internal clocks have been timed with care. The human body clock, for example, has a period close to 25 hours. An hour's discrepancy is only 4 percent of 24 hours, a seemingly reasonable tolerance. This close adherence to the period of the earth's days has prompted the name *circadian* for this class of biological clocks, from the Latin *circa* (about) and *dies* (a day).

An hour in 24 seems negligible, but like the difference in speed between cars on a racetrack, it has rapidly cumulative effects. Some cars will fall behind and others will pass, even repeatedly, on a circular track. But living organisms are not racing the earth; what is needed is synchrony, so corrective measures must be taken to maintain it. If your inaccurate clock could not be reset, then you would have to travel continuously to stay in synchrony with your surroundings (40 miles an hour at the equator, to compensate for the mismatch of an hour a day). If the clock were dead accurate, you would be rooted in the time zone of its manufacture, the zone of your birth, or even perhaps in the time zone of your mother's birth, and of her mother's! Moreover, this perfect biological clock would have to be immune to every resetting influence: if it slowed down in cold weather, or jumped ahead or back in reaction to shocks, or stopped while the batteries were being recharged, or just gained a second every day, it would not long be useful. Nobody wants a clock like that, including Nature.

Without the means to reset their 25-hour internal clock—a capacity that many sightless individuals and some with normal vision seem to have lost—nontravelers would drift in and out of step with the 24-hour world. If the discrepancy remained always an hour in 24, then synchrony would recur fleetingly every 24 days. Improving the internal clock's match to the external period does not eliminate the problem; it merely slows the inevitable drifting in and out of synchrony. With a discrepancy of only 1 minute in 24 hours, an imperturbable internal clock would drift through the same cycle, regaining exact synchrony once in every 1440 days. To maintain synchrony between our internal clock and the earth's rotation requires more than a close match between the two periods: it requires some cue, a cadence caller by which our clocks may be reset daily to agree with local time where we live.

This capacity for phase resetting is the essence of any biological clock's utility. By resetting on cue, it keeps our insides in the right time zone, in phase with local time, adjusting for any mismatch due to travel east or west and the discrepancy between our clock's period and the earth's.

But what is the cue? Not much can yet be said with assurance about humans. But about every other organism that has been examined, the answer is clear, and might have been guessed. The signal must be inflexibly connected to the rotation of the earth, perfectly dependable in its daily appearance, and utterly unmistakable to the organism. One good choice would be light, detected with such sensitivity that no amount of overcast could disguise the hour of sunrise. And most species' internal clocks do indeed react more sensitively to light as a timing cue than to anything else. Continuous illumination at the intensity of moonlight, in fact, is enough to arrest the progress of the circadian cycle in fruit flies and fungi in my laboratory. The clocks in mammals are not that sensitive, and in humans they seem very much less so, but for almost all species tested, watching for light is the best way to learn what time it is.

Phase resetting of circadian clocks is the main theme of this book. The story of these clocks inevitably melds into the study of dynamical systems: of chemical oscillators, of the equations governing unstable equilibria, of the geometry of cycles in multidimensional state spaces (see Chapter 8). To understand body clocks, however, it is not necessary first to have a course in geometrical dynamics. The same phenomena that drive our bodies' daily rhythms can be seen acting in single cells, and much of what is securely known and apparently universal about biological clocks concerns only timing relationships. The geometrical principles involved are plainly laid out on the earth's surface.

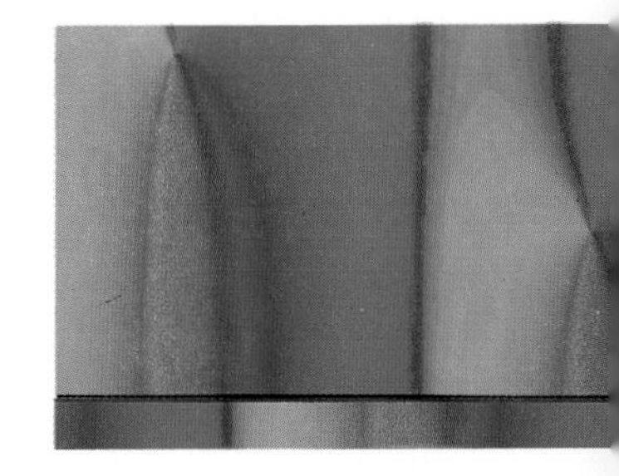

Patterns of Timing in Space

A curious principle of timing that turns up everywhere in biological contexts seems to have escaped even the prescient pen of Lewis Carroll.

The modern astronaut evades jet lag by neglecting to continually reset his wristwatch and his body clocks. Should he attempt to reset, he would still excuse himself from what seemed a bewildering conundrum to Pigafetta by glib reference to the international date line. If you must reset, then you need only accept the line, and some problems go away. Yet there remains something peculiar about this line. What kind of line is it? Evidently it is not a circle since it was encountered only once in a flight from Hawaii to Hawaii. So it ends somewhere. What if I go to the end and walk around it, cross over, walk around the end and cross over again, and so on? Can I ratchet myself into the remote future or past, a day at a time? Unfortunately not. The international date line has its endpoints at the poles, the two mirror-image points where the boundaries of all the time zones converge. The time zones mark different phases of local time, and the orderly convergence of all time zones and phases marks a *phase singularity.* In walking around an endpoint of the date line, I also walk backwards around the phase singularity: what is done in stepping across the line is undone in walking around its endpoint.

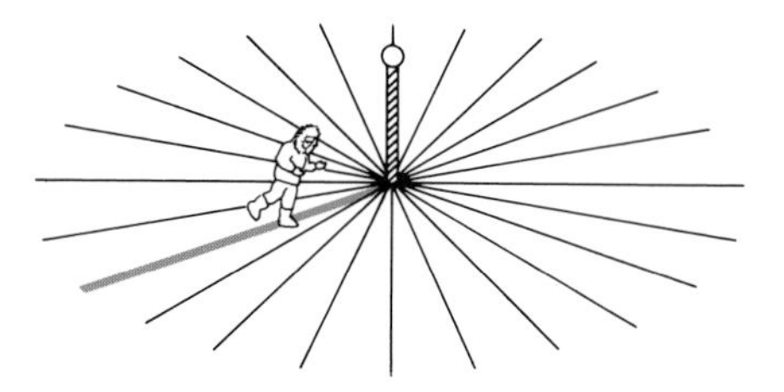

Since the date line has an endpoint, it might seem possible to cross the line repeatedly into the future by sneaking around the end. This paradox can be avoided only if a phase singularity lurks in the endpoint: as it does, shown here by the converging time zones.

Is there perhaps some way of drawing the boundaries of time zones to avoid phase singularities? We might try by adopting another definition of *time zone.* The conventional time zones of cartography correspond closely to the time zones of city streetlight operation only on the equinoxes. At any other time of year there is a disk of darkness surrounding one pole and a disk of daylight surrounding the other, due to the tilt of the earth's axis that gives us our latitude-dependent seasons. When measured from the sun's daily maximum elevation, the time zones of convention do indeed have a phase singularity at the poles all the time, but the time zones of experienced sunrise or sunset terminate tangentially along the singular fringes of those seasonally expanding and contracting disks; they have a singular *point* only on each equinox. This attempt to evade the necessity of a singularity has only enlarged it! For topological reasons, convergence is inevitable, and a singular *point* is its minimum manifestation.

Time zones cannot be organized in arbitrary ways. The equator describes a closed-ring path in space along which the time zones must run through a full cycle, so that if you fly along the equator from one time zone to the next until you return to where you started, you pass through a full day. Your path cuts the globe into two territories, one hemisphere to the north and one to the south. Along that path, the equator, all the time zone lines enter each hemisphere and they don't come back out. So no matter how the zone boundaries may be drawn, somewhere inside they must end or come together. Whether or not the time zones converge neatly to a mere point, somewhere in each hemisphere the systematic cyclic pattern of phases must break down. There the time is ambiguous, indeterminate. What happens to a clock at such a place? Nothing, of course; it will keep its own time in violation of the pattern, for a pattern cannot be adhered to if it requires that a clock must indicate no definite time at all. (A confusion about this point resulted in circadian clock experiments being conducted at the South Pole in 1962 to see whether anything peculiar would happen. Nothing did.) We shall later examine phase singularities in which the clock *must* adhere to the pattern; its only choices then are to abandon the usual cycle of phases or to simply quit altogether.

The ultimate basis of this dilemma of clocks is geometric: it stems from the fact that longitude, like compass direction and periodic time, is represented by points on a circle, with no point on the circle distinguished as the beginning/end except by arbitrary convention. This feature contrasts starkly with the geometric underpinnings of the numbers we are compelled to use, so clumsily and inappropriately, to describe compass direction or periodic time or longitude. Every number represents a point on a line. To get from one number to another you must go either higher or lower: you can't take your choice as you can on a circle. If you keep going in one direction on a line, you will never get back to where you started, though you must on a circle. The numerical jump on a compass (360 degrees jumps to 0 degrees), a clock (12 or 24 hours jumps to 0), or a globe (−180 degrees jumps to +180 degrees at the date line) is a consequence of trying to adapt numbers for a purpose alien to their nature. This distinction between a circle and a line remains no matter how they may be distorted, so long as no cuts are made and no parts are glued together. The branch of geometry that deals with such properties is called topology, and the topological point of view unifies much of the study of biological clocks.

Color-coding Periodic Time

Another familiar quantity that takes its values on a circle is color, or more specifically one of the three qualities that make up any color: its hue (we neglect the other two attributes of color, its saturation and brilliance). Hues have names and an ordering: yellow is close to green is close to blue is close to violet is close to purple is close to red is close to

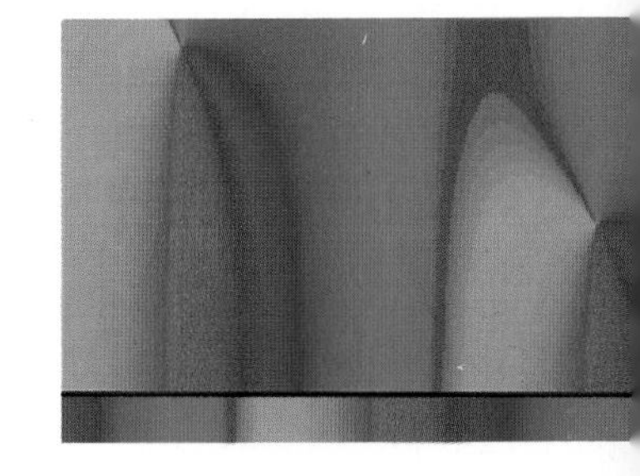

orange is close to yellow is close to green is close to blue is close to violet is close to purple is close to red, and so on. We all learned that in grammar school; it is the "color wheel" lesson.

The physicist's spectrum does not close in a cycle. It reaches only from red through blue and has no "purple." The sensation of purple is elicited by no pure spectral color, but only by a mixture of the extremes, red and blue. It is not between them in the same sense as yellow is between them. Yet, going from blue through violet and purple to red, it is possible to finish the circuit without backing up or repeating a color, to back to where you started. The circle of perceived hues, as opposed to the visible range of spectral frequencies, thus constitutes a natural language for talking about periodic, cyclical things that must come back on themselves: angles about a point, time zones, phase in a cyclic process. Color is a natural language for describing the state of such systems. It will be our language for straightforwardly saying simple things about cycles that sound like riddles in the more familiar language of numbers.

The mere fact that our perception of hue is ordered on a ring implies inescapably that we must also be capable of a hueless sensation, a totally unsaturated color of ambiguous hue. We are indeed, and it is no big deal: it is just the series of achromatic "colors" white–gray–black (according to intensity). This sensation conveniently embodies the idea of a phase singularity being assignable to no point on the circle but at the same time being arbitrarily close to grays tinged with every hue on the circle.

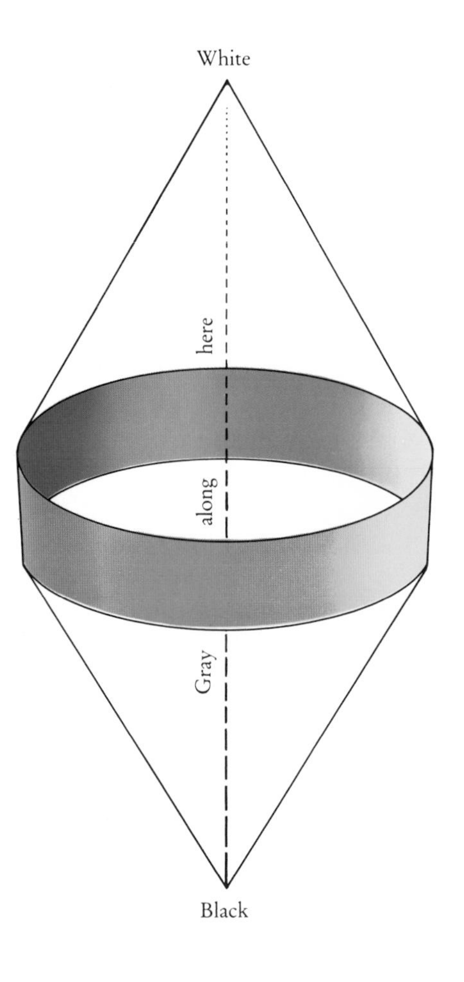

The sensation of color has three independent attributes, which can be represented conveniently as a *color solid*. Brilliance is indicated here along a vertical axis between white and black; hue varies like longitude along a circular axis; and saturation decreases from the surface toward the white-black axis of achromatic grays. Colors of the same brilliance and saturation then appear as the familiar color wheel. The sensations evoked by the physicist's spectrum lie mostly along an equatorial arc of saturated colors stretching from red to blue.

Time Zones for the Tides

This book is about occasions when periodic timing is annihilated, about what happens when a biological clock is confronted with a singularity in its internal pattern of phases. Our understanding will advance by progressively generalizing the idea of time zones. As a first step away from the rigid simplicities of drawing meridian lines on a rotating ball, let us consider another kind of cycle with time zones and singularities.

Land-dwellers usually care a lot about sunrise and sunset. Living organisms mark time in cycles of 24 hours in step with the alternation of dark with light needed for vision or photosynthesis. To some organisms, the moon is as important as the sun, but not for light: concern for the rise and fall of the tides can dominate daily life as much as the alternation of light and darkness. You might feel this way about things if

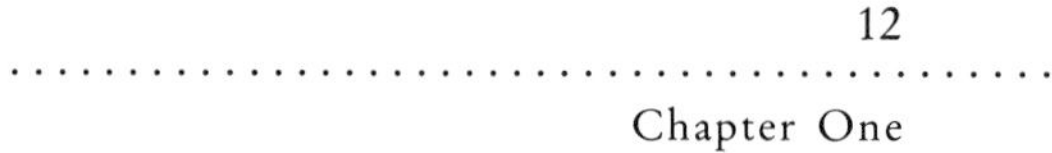

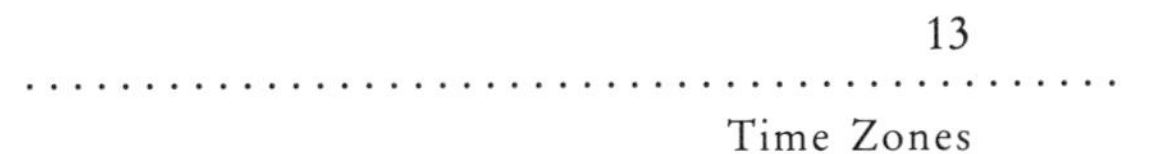

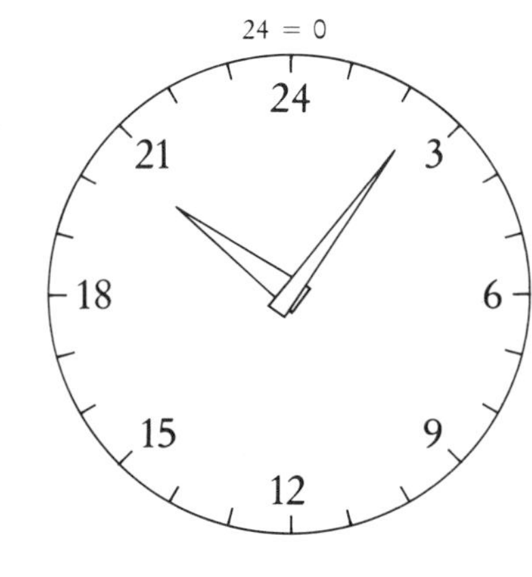

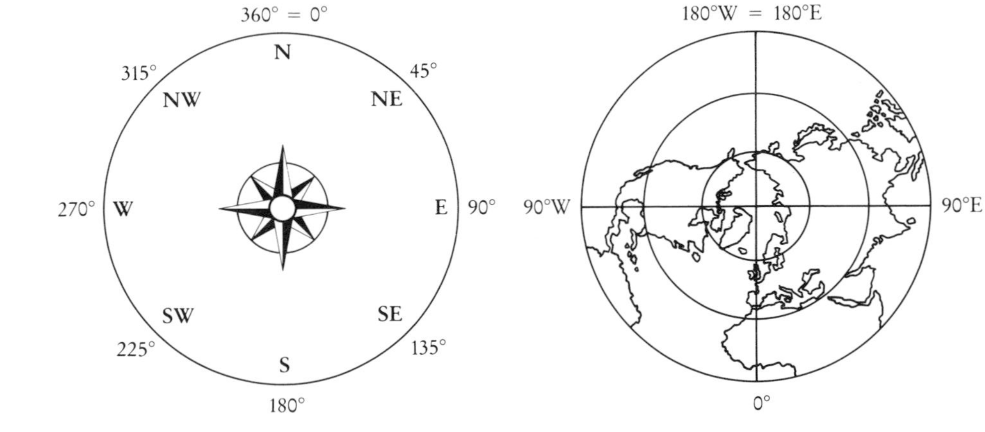

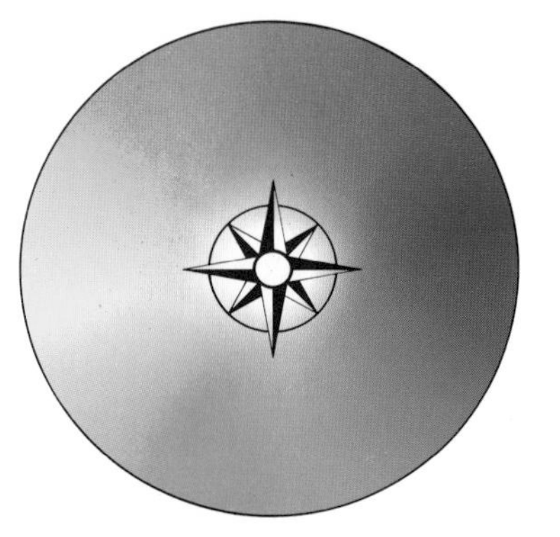

Compass card, clockface, and globe in polar projection all conventionally indicate a circular measure by unfortunate analogy to the real numbers. Perceptual hue smoothly labels the same quantities without introducing a discontinuity at 0 = 360 degrees, 24 = 0 o'clock, or 180°E = 180°W longitude.

you were an intertidal crab, for example, or anyone else living in the marshy grasslands of the southeastern United States, or if you operate a fishing boat in shallow water, or even if you like to run on the beach every day. In fact, if you were to habitually start your day by watching the moon set, you would be following the innate period of your body clock much better than you do by sticking to a schedule based on sunrise (more on this later).

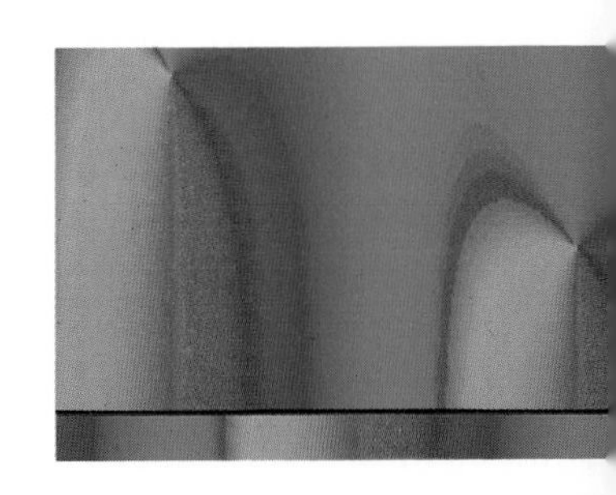

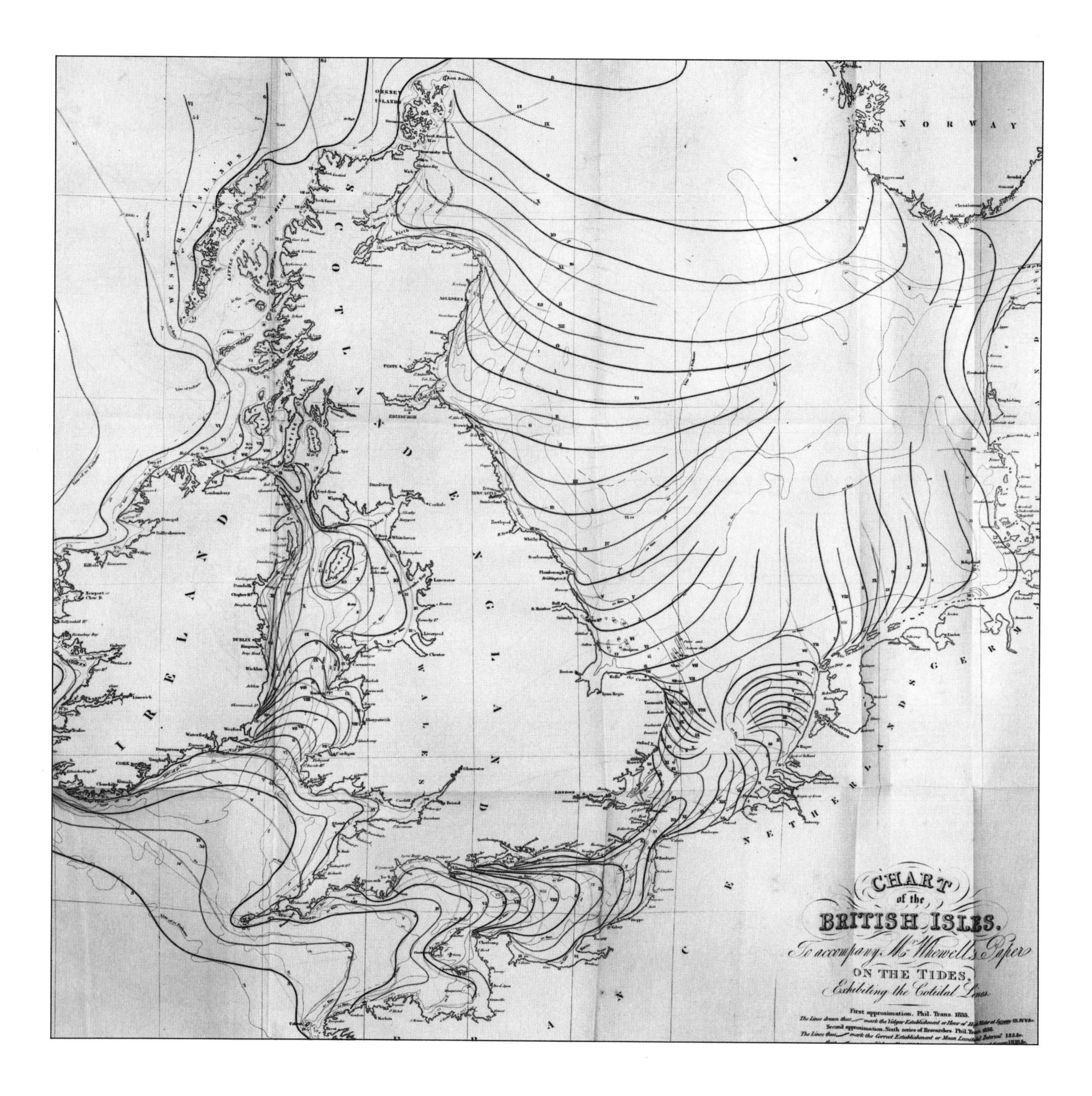
NORWAY
SCOTLAND
IRELAND
ENGLAND
WALES
WESTERN ISLANDS
ORKNEY ISLANDS
EDINBURGH
DUBLIN
LONDON
CHART of the BRITISH ISLES,
To accompany Mr Whewell's Paper
ON THE TIDES,
Exhibiting the Cotidal Lines.
First approximation. Phil. Trans. 1833.

The tides are governed by both moon and sun, their relative proportions of influence varying from place to place. The moon-dominated tides roughly repeat every 12 hours and 26 minutes (almost an hour later every day). But their timing varies from place to place. Francis Bacon, near the end of the sixteenth century, had first framed the natural question: What is the pattern of timing of tidal crests? They can't be simultaneous everywhere if the oceans have only a fixed volume of water. By the early nineteenth century, there were plenty of tidal recording stations along the shores of all the world's oceans and on a few islands. In the early 1800s William Whewell (who also gave us the word *physics*) organized what may have been the first international geophysical collaboration to sketch on a chart of the North Sea a series of "cotidal lines," as he called them: the curves along which you find high tide now, the curve along which you'll find it an hour from now, two hours from now, and so on.

Facing page: Times of the tidal crests along the coasts of the North Sea were used to interpolate timing at sea between London and Amsterdam. The cotidal contours were found to converge to a point around which the pattern rotates as the tide rises and falls at any point along the coast. At the convergence point, any fluctuations in the depth of water (for example, due to the weather) have no relation to the period of the semidiurnal M_2 tide at shore stations. There is a second such point, also with counterclockwise rotation, further out to sea to the north.

Because there are no cliffs of saltwater in the ocean, you might imagine that you could trace cotidal contours smoothly across the sea to match shoreline tidal data. And you can, up to a point, but then an unforeseen difficulty emerges: in a few strange places the contours converge to a point! As Whewell put it,[3]

> It appears that we may best combine all the facts into a consistent scheme by dividing [the North Sea] into two rotary systems of tide-waves;—[in each] space the cotidal lines may be supposed to revolve round [a point] where there is no tide, for it is clear that at a point where all the cotidal lines meet, it is high water equally at all hours, that is, the tide vanishes.

That sounds like time zones and phase singularities again: the tides have their own kind of North and South poles, singularities where all the tidal time zones meet. The cotidal contours converge in clockwise or counterclockwise rotating circular order, much as the time zone boundaries converge to the opposite poles. At those points the lunar rhythm vanishes: the depth varies irregularly with the weather or follows some other frequency-component of the tide, with a daily period, for example. A moderately detailed map of the entire globe shows about eight major clockwise and eight major counterclockwise convergences of the dominant component, the lunar M_2 tide, as it is called.

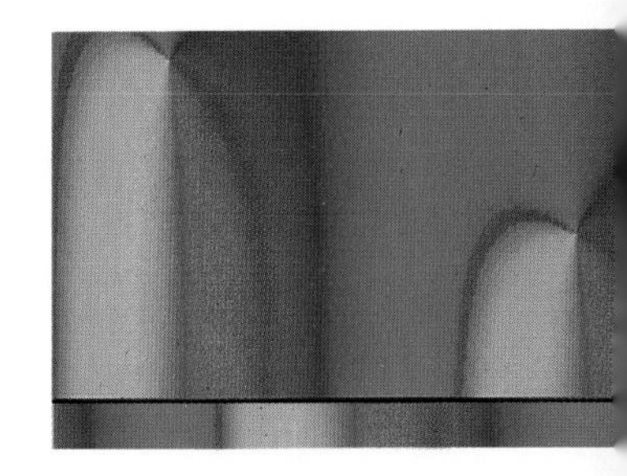

Dubbed *amphidromic points,* or *amphidromes,* these phase singularities organize the spatial pattern of tidal timing throughout the globe. It is not easy to grasp intuitively how and why they arise where they do,

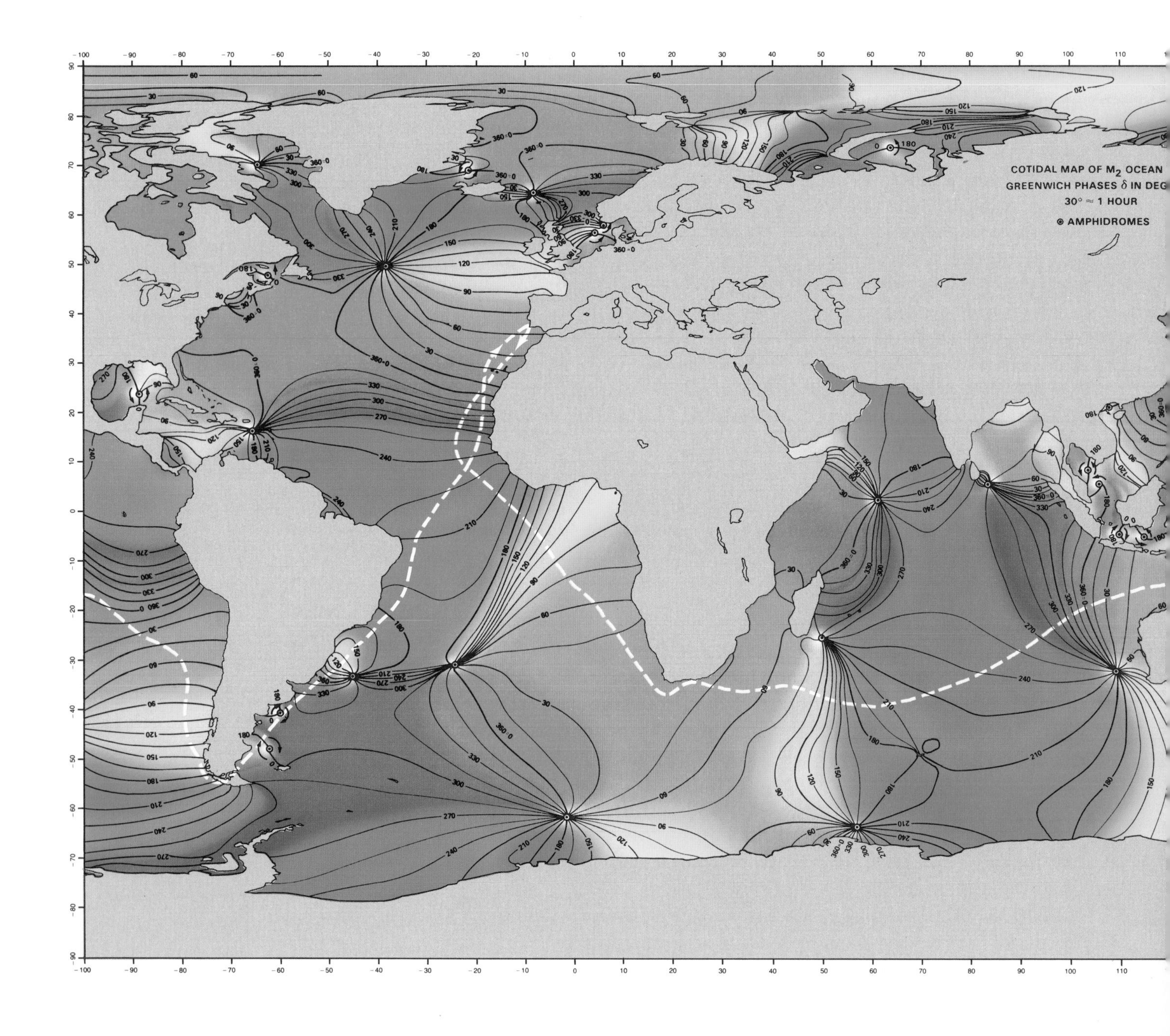
COTIDAL MAP OF M2 OCEAN
GREENWICH PHASES δ IN DEG
30° ≈ 1 HOUR
AMPHIDROMES

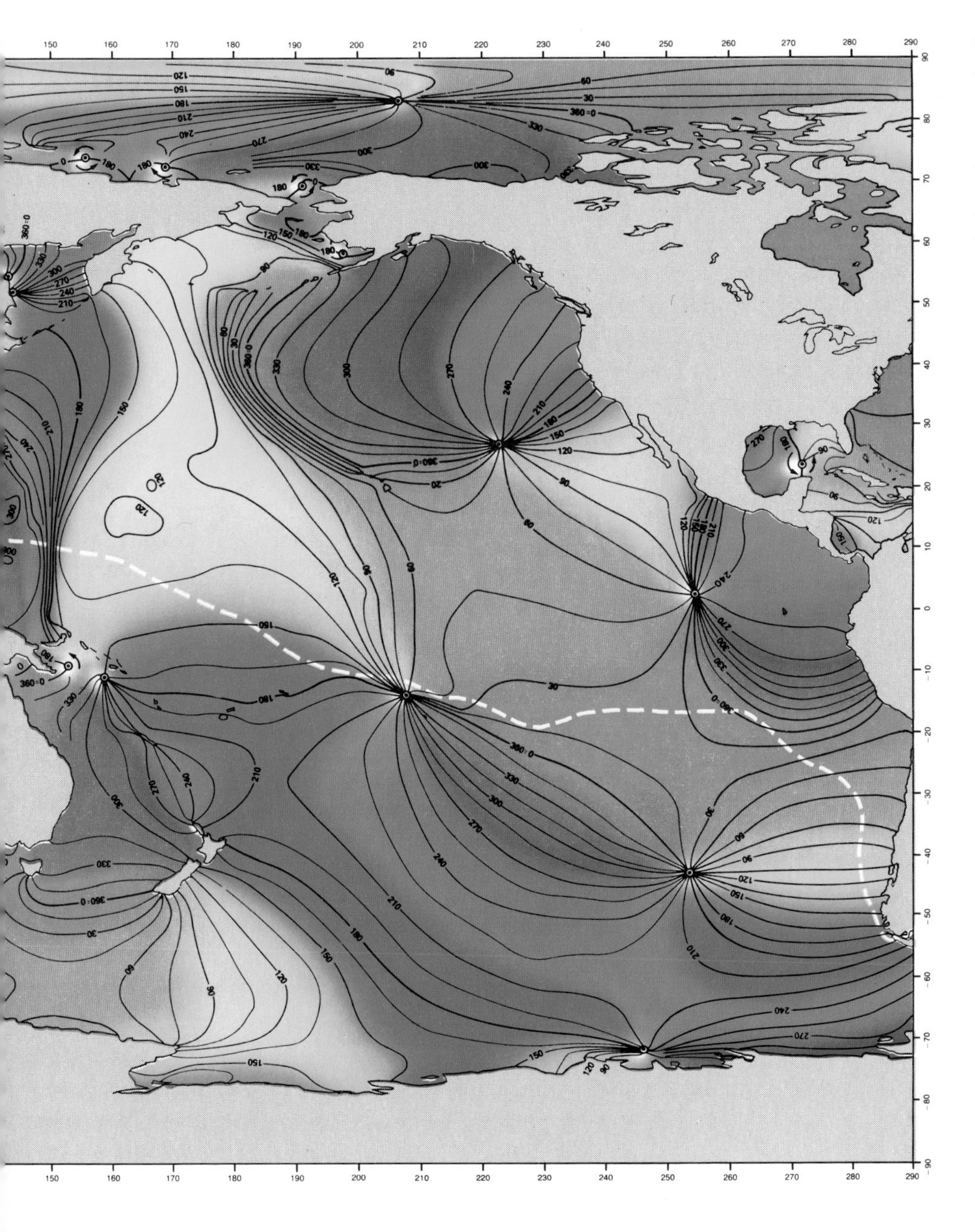

This cotidal chart encompasses all the world's oceans. About equal numbers of clockwise and counterclockwise amphidromic points appear; the pairing would be exact only if the seas also covered the land masses and if the mapping were done in fine detail. The white line shows the route of Magellan's ship.

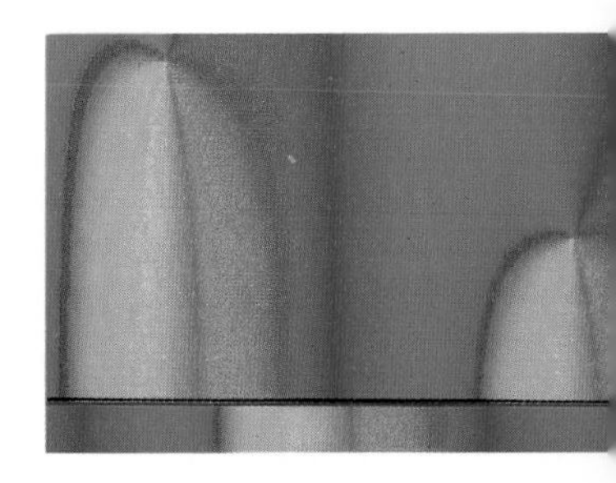

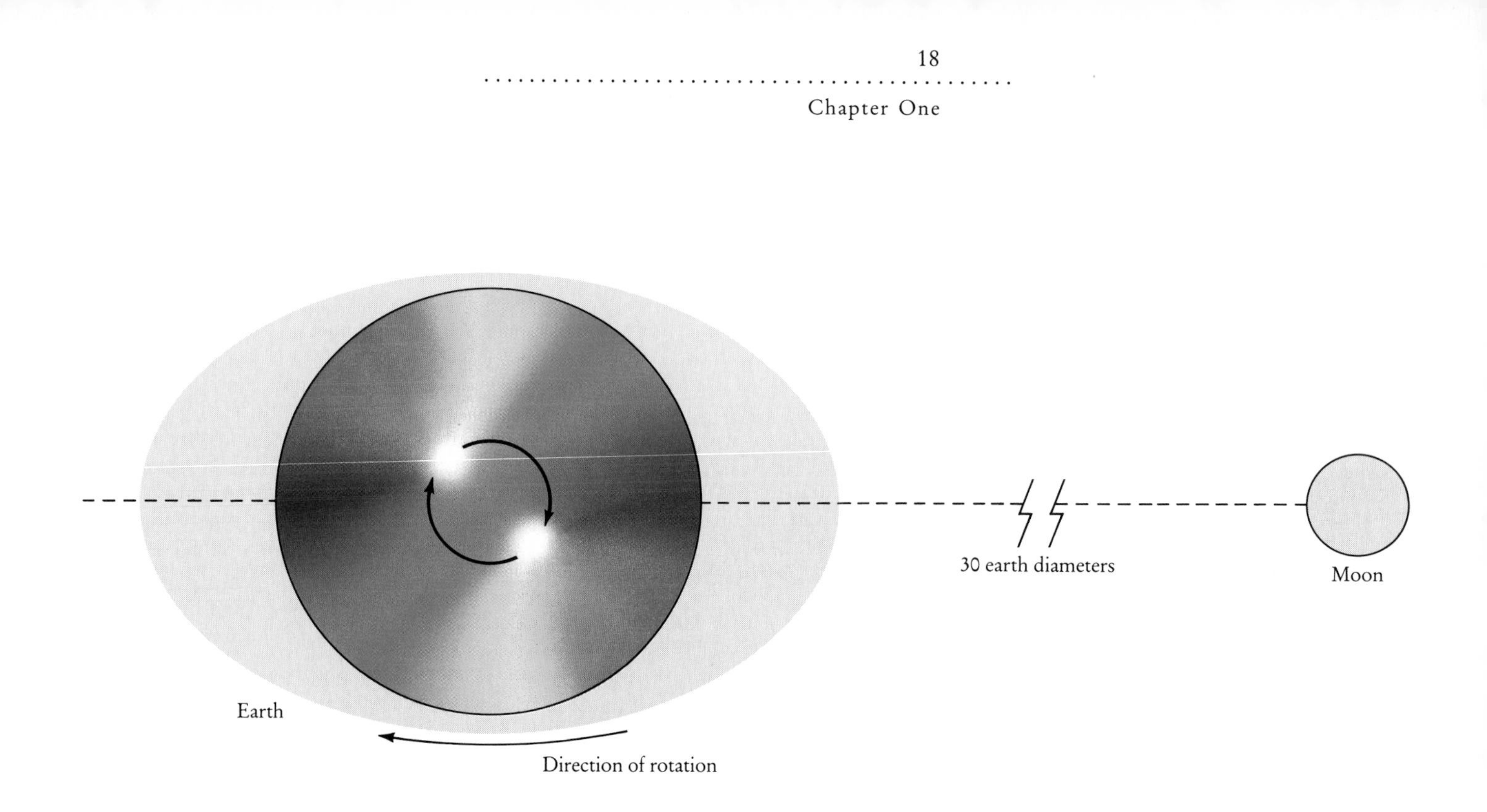

This view of the earth and moon from high above the South Pole is highly idealized, for example, by deleting continents, neglecting the 23-degree tilt of earth's rotation axis, and placing the tidal bulge directly under (and opposite) the moon rather than several hours lagging. However, the twofold amphidrome that would occur at the pole in that symmetric world is here shown more realistically decomposed into a moving (geographically stationary) pair of familiar onefold amphidromes. A mirror-image pair belong near the North Pole. The colored map on pages 16 and 17 shows that there are in fact continents and many additional pairs. Purple here indicates high tide and yellow low tide.

how they move seasonally, and so on. But the underlying physical principles are known with precision. To begin with, at least two of each kind must exist for the simple reason that the lunar tides resemble a pair of diametrically opposed gigantic bulges in the world ocean, held in place by the moon while the earth spins within. On a perfectly symmetric globe with an upright axis there would be a neutral, amphidromic point at the North Pole and a mirror-image point at the South Pole. Each hemisphere would have *two* contours of maximum tide, rotating clockwise around the North Pole and counterclockwise around the South Pole. But in fact, the earth's axis is tilted and its oceans are shallow and irregular, so the symmetry is spoiled. This does not eliminate the topological necessities. There must still be centers for a twofold clockwise rotation and a twofold counterclockwise rotation, although they could be manifest as four simple centers. Might the many additional mirror-pairs have arisen like eddies in our resonantly sloshing seas? Might some also be hidden within the borders of continental landmasses? Answers can be calculated in surprisingly accurate detail by taking into account the inertia and viscosity of water, the shapes of continents, and the slopes of seabeds along with some tidal observations. In fact, the map on the preceding page represents such a computation.[4]

The track of Magellan's epic voyage is marked in white on the map on pages 16 and 17; completely unaware, he traversed the basins of several major amphidromes in the world's oceans. Because there are (including relatively minor ones) 10 clockwise and 9 counterclockwise amphidromic points south of his path, Antonio Pigafetta and colleagues experienced one extra cycle of the tides. But Pigafetta was not monitoring the tides and never noticed the extra ebb and flow.

He *was* monitoring the day/night cycle. Because of the single convergence of time zones south of the ship's course, the returning crew missed out exactly one complete cycle of day and night. Why then did they not notice it? A modern traveler would notice the progressive loss of major fractions of a day long before reaching home again because he would be carrying a timepiece locked to his original time zone. But Magellan's men carried only their biological clocks, and biological clocks have evolved for resettability. The first large-scale experiment probing the nature of these clocks may have been unwittingly conducted by Antonio Pigafetta and his shipmates while struggling westward back to Portugal through disastrous misadventures on the far side of the world. As they sailed around the world their biological clocks were reset day by day to show no discrepancy with local time. Imperceptibly, this systematic resetting brought them closer and closer to its culmination in Pigafetta's Paradox: the unnoticed vanishing of a full day from their calendar.

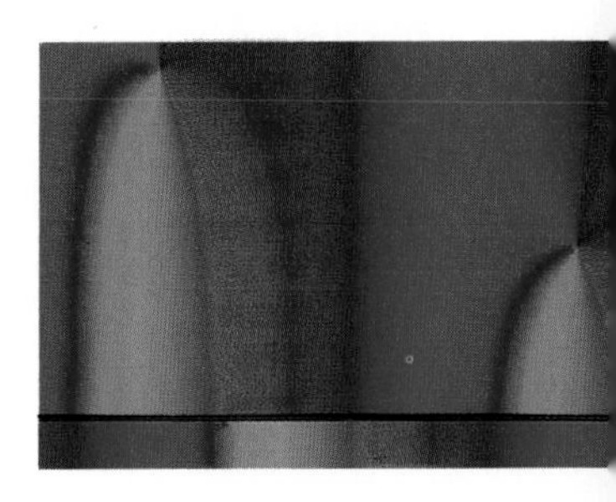

In each time zone the daily rising of the sun synchronizes internal clocks throughout the ecological community.

Chapter Two

The Ebb and Flow of Consciousness

Much wonder groweth upon me, yea astonishment seizeth me. [Men go abroad] to wonder at the height of mountains, at the huge waves of the sea, at the long courses of rivers, at the vast compass of the oceans, at the circular motion of the stars; and they pass by themselves without wondering.

ST. AUGUSTINE, 354–430
CONFESSIONS

Nothing could be more familiar than the alternation of sleep and waking, the relentless pattern of our days and nights. As the earth turns, the texture of our consciousness undergoes its own revolutions, from the unrecorded fantasies of solitary dreaming, to the collective fantasy of daily social and commercial life, then back to solitary dreaming. Anyone who has tried to escape from this regime of clocklike alternations knows that it is not strictly enforced by the alternation of light and darkness; and knows also that it cannot long be evaded. The timing of that geophysical clock is built into us, and not just by lifelong habit of submission to the global rhythm: it is part of our physiological inheritance. Under ordinary conditions, the cycling of this innate biological clock is synchronized by the overwhelmingly greater geophysical clock on which it is modeled. But the inner clock still beats and plays a vital role in our daily lives. Our task here is to listen for the inner beating of the biological clock in those rare situations where it can be heard independently.

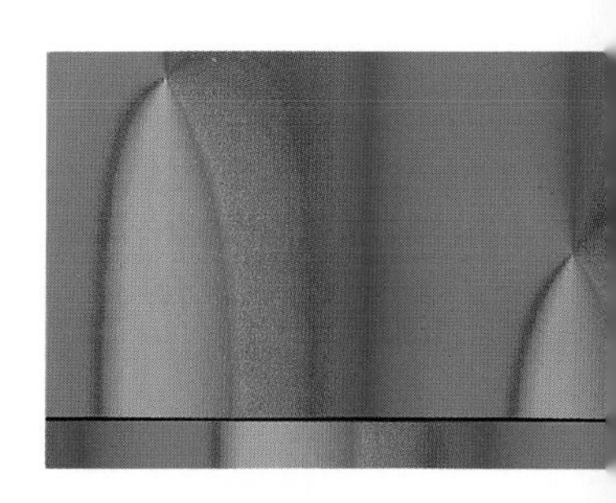

Temporal Isolation

Doorway of the Aschoff/Wever temporal isolation bunker in Andechs, Bavaria.

Childrens' voices from the 800-year-old monastery of Andechs, in Bavaria, ring through the tall pines flanking our wooded path. Beyond the woods, the path opens into a short road; Rutger Wever's house stands quiet in the chilly October sunlight, and a little further along, we can see the Max-Planck Institute. In between, a mound to the right of the road invites entry through a tall slot of concrete like a bunker doorway, half black in cold shadow. Comfortably warm inside and unaware of our presence in the vestibule, a woman meditates beneath this artificial hillside, in Wever's *temporal isolation facility,* an efficiency apartment without windows, television, or any other external cues to the time of day. But there are plenty of books inside, and a newspaper is provided once a day when the subject sleeps.

Right now, at the beginning of a month-long experiment, her habits are being observed during exposure to a normal 24-hour light cycle: the lights are turned off every night at 11:00, and come on again at 7:00 in the morning, synchronized to civil time outside. Except for a small reading lamp considered too dim to matter, the subject has no control of the lights. Her core body temperature is continuously recorded by an anal thermistor as it rises in the daytime and falls at night through a 1-degree (C) range. Urine is collected at intervals to measure water and electrolyte balance. Microswitches also monitor her activity and record her sleeps, whether at her desk or in bed. All of her monitored quantities exhibit rhythmic ups and downs on a 24-hour schedule; the sleep/activity cycle is only the most conspicuous among them. It is worthwhile to examine it with care.

Imagine a long black tape emerging from the chart recorder at a speed of one centimeter per hour, painted white when the subject is active. During a normal 24-hour cycle, a white segment about 16 hours (16 centimeters) long begins every 24 hours when she wakes up. To display this long record in a way that brings out its rhythmicity, paste consecutive 24-hour segments horizontally, one above the next, so that activity times make a vertical column of white segments alongside a black vertical column of sleep. By analogy to the same method of painting a picture from a time series in television technology, this display is called a *raster.* The television raster is the stack of 525 horizontal lines painted across the screen one after the other, each in less than 100 microseconds, rather than 24 hours, to construct an image. Just as on the television screen, in our display of alternating intervals of sleep and waking, the

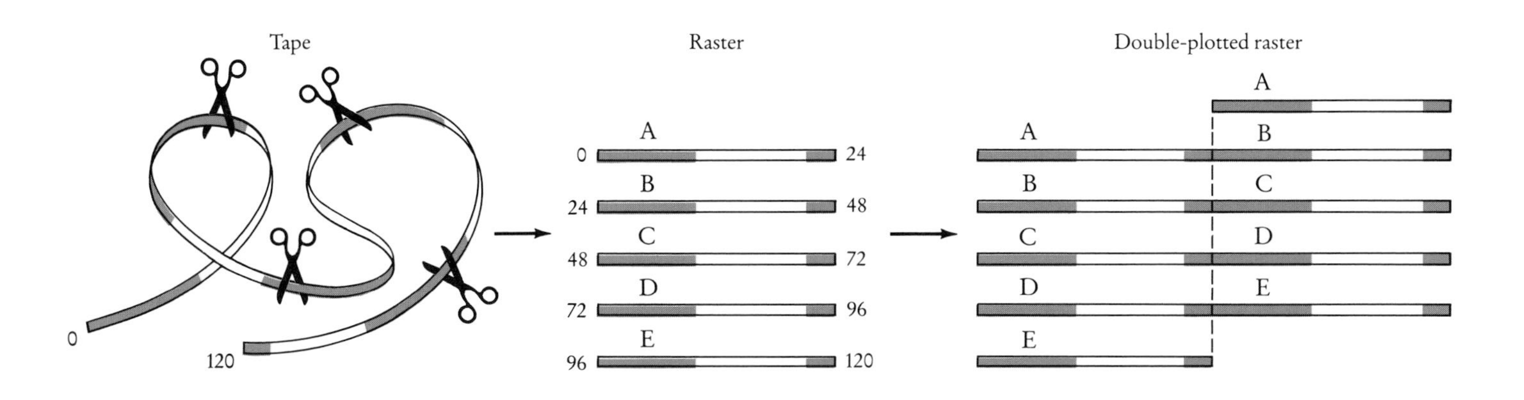

A long tape bearing the marks of a rhythm is cut up into equal segments, ideally of period exactly equal to that of the rhythm. Here we take 24 hours as that period. The segments are then stacked up so that successive cycles appear one atop the other. But now a spurious discontinuity has been introduced at each time when the scissors separated one segment from the next. Continuity is restored by double-plotting, duplicating each segment to appear also to the right of its predecessor.

end of each line is the beginning of the next line down. Continuity across the artificial line break is restored if we duplicate the entire stack of segments and paste it just to the right, shifted up one line. Each day then continues smoothly into the next from left to right, and each day is plotted both to the right of and below its predecessor. This is called *double-plotting.* The double-plotted raster shows four vertical columns of alternating black and white because the long data tape has been cut up into stacked segments of length exactly equal to the period of the activity cycle: 24 hours in this case.

Suppose control of the isolated apartment's lighting is now handed over to the subject. A typical subject will continue his or her body-temperature rhythm and sleep/wake alternation as before, but they will drift later by about an hour each day. In this 25-hour circadian rhythm, the column of black bars drifts right as it descends at constant slope, shifting across another 24 hours every 25 external days, that is, every 24 internal ones. If a newspaper is still provided "once a day" during the subject's sleep interval, then a time must come—after about 25 external days—when today's paper arrives at the lab before yesterday's has been given to the temporally isolated reader.

The double-plotted raster on the next page shows a 24-year-old man's reversion to the innate 25-hour rhythm when he was excused for six months from strict adherence to worldly rhythms. Such records of the timing of unforced human activity continue to surprise everyone who studies them and, most astonishing of all, to surprise the subject who generates them. "Naps" perceived as half an hour long are commonly full-blown sleeps of 8 hours or more: it seems we are utterly incapable

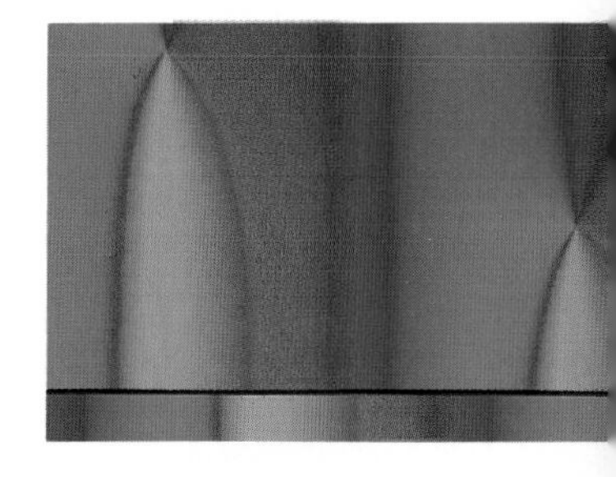

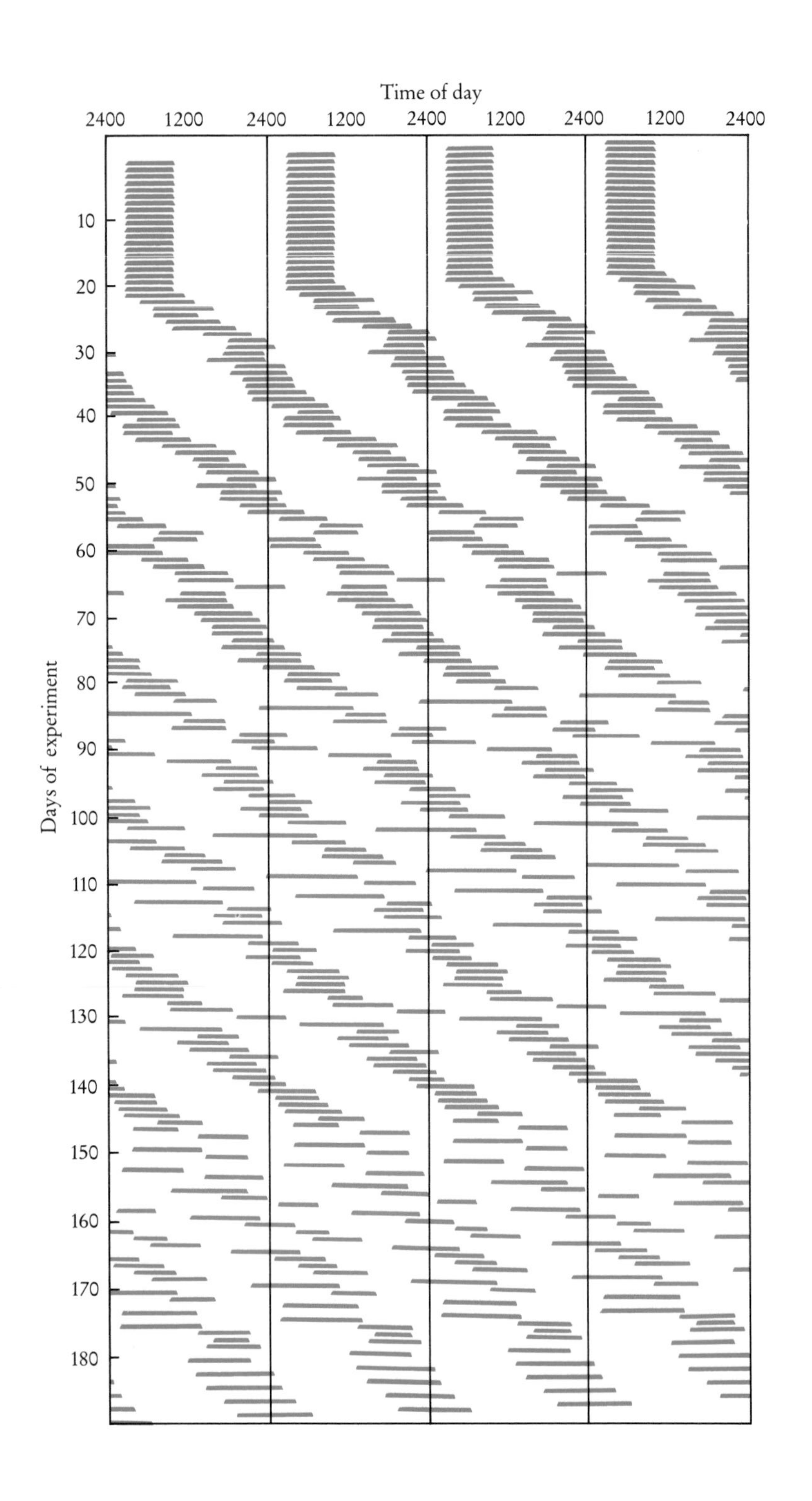

In double-plotted raster format, a human sleep/activity rhythm (in blue) adheres to the 24-hour cycle of light and darkness (top). It then continues with its own native period near 25 hours when the subject is allowed to operate the lights himself. Because successive lines of the raster are 24 hours long, the 25-hour rhythm appears to drift downward to the right by an hour from one line to the next. Were segments cut at 25-hour intervals, this rhythm would stack up vertically, and the 24-hour rhythm above would slant to the left as time increases downward. The color raster plot shows six months of sleep and waking by a temporally isolated young man, quadruply plotted.

of gauging elapsed time during sleep. No less appalling is our inability to gauge even the length of day. People in temporal isolation for at least a few weeks commonly switch from a schedule of about 8 hours sleep and 17 hours activity to sleeps and intervening active intervals irregularly about half again as long, or even regularly just twice as long, as seen at the end of the six-month record on the facing page. In one of the first experiments, the subject's average time of bed rest without getting up was 19 hours. This behavior alarmed the investigators—repeated sleep of such duration seemed very unusual in a healthy man. But people take these marathon sleeps and work shifts comfortably in stride and don't even notice that some sleeps are as short as 4 hours and others as long as 18 hours, or that some "daytimes" continue as long as 30 hours without a sleep.

During such complex vicissitudes of sleep and activity, the body temperature rhythm adheres steadfastly to its 25-hour period. Jurgen Aschoff, director of the Max-Planck Institute in Andechs, Bavaria, was the first to report the internal temperature rhythm, thus revealing the faithful regularity of an internal circadian clock even while some aspects of behavior fail to keep pace with it.[1] Aschoff found that in particularly long sleep/wake cycles the subject's core temperature may wax and wane twice in one cycle. The subject continues to accept the next newspaper without question at each breakfast, while he or she continues to take three meals (somewhat larger, while losing a little weight) per "day," noticing no particular increment of daily accomplishments. The subject's confusion and skepticism were often dramatic when, after two or three weeks by private count of "days," the supervising scientist entered to announce that the agreed-upon month was finished—but could not convince the subject until he produced a stack of "future" newspapers to prove it.

Longer histories of sleep and waking have been recorded for ground squirrels and hamsters living for years on end in solitary confinement without time cues. Their inner clocks seem to synchronize sleep and waking more reliably than ours do. However, these small mammals have revealed gradual changes in the length of the circadian cycle by a few minutes over the course of years. There are hints of similar behavior in humans, but there are as yet no such long human records. Six months is about the longest anyone has cared to remain in temporal and social isolation up to now—though we may have missed some spectacularly prolonged experiments that went unrecorded centuries ago in the timeless darkness of castle dungeons.

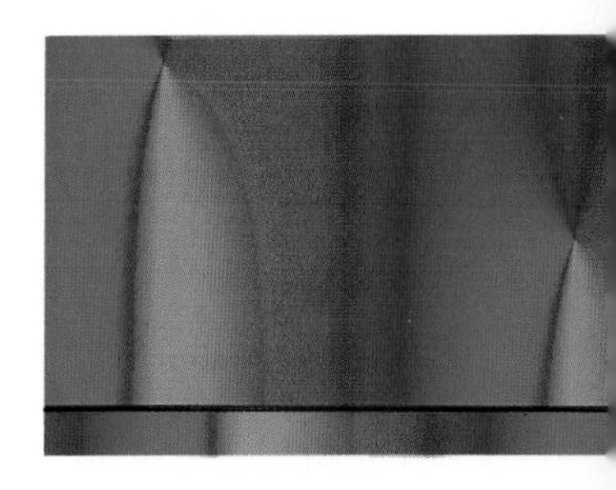

BIOLOGICAL CLOCKS IN MEDICINE AND PSYCHIATRY

In the usual civil day/night cycle, everything is constrained to a 24-hour repeating pattern, so our physiological functions wax and wane in concert with the more conspicuous regularity of sleep and waking. Implications for the practice of medicine are now obvious,[2] but it takes decades to reduce new insights to clinical practice. An example with immediate practical appeal is the daily rise and fall of the pain threshold in your teeth. Its timing is such that we ought to visit the dentist in the afternoon, or to avoid his services altogether in daylight hours if we have just arrived in Chicago from Japan. The effectiveness of anesthesia shows a comparable high tide in early afternoon; the dose required in the morning might be excessive later

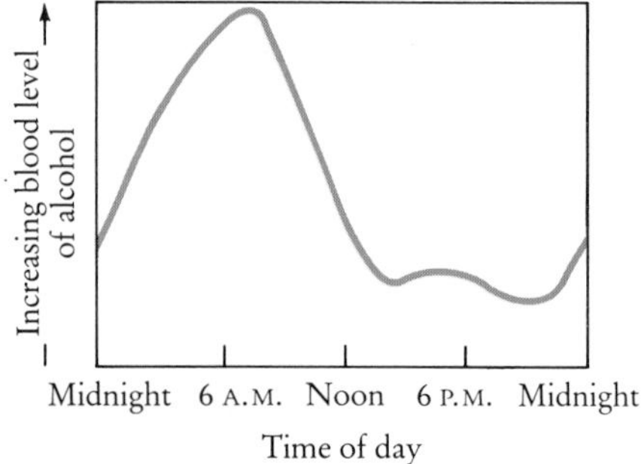

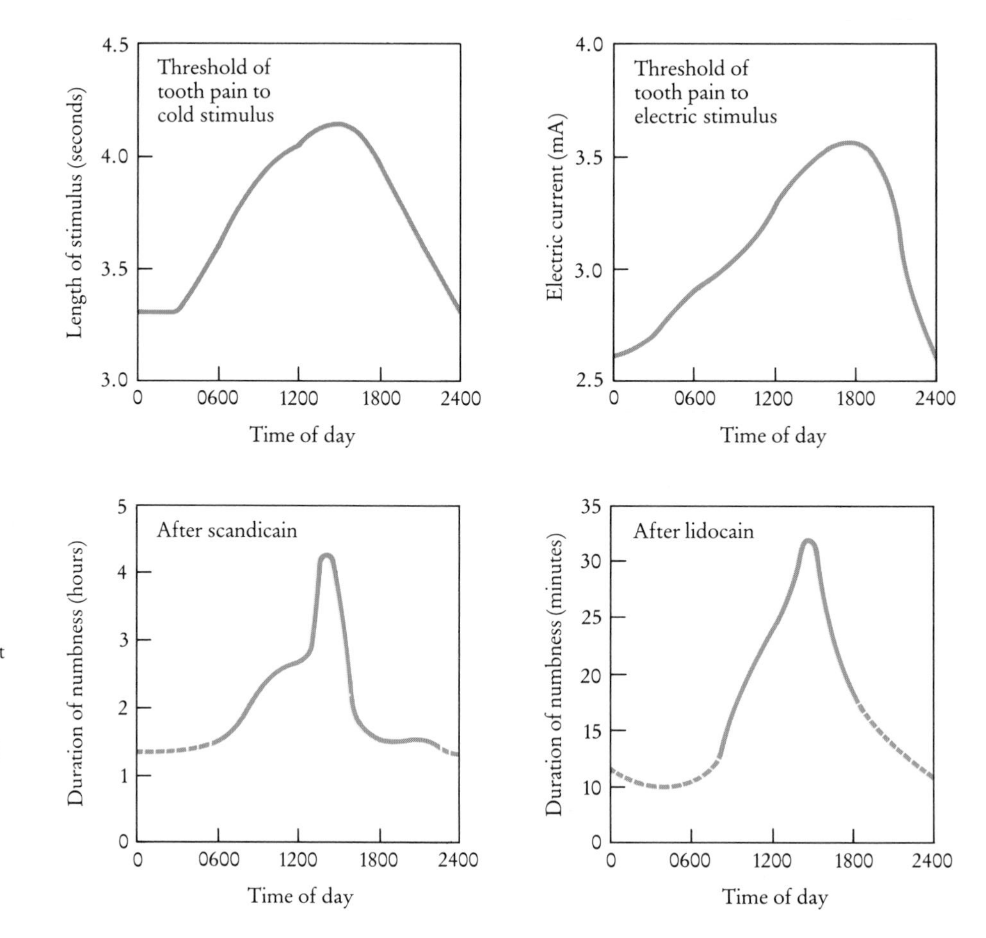

In the afternoon, dental pain threshold is higher by half and numbness from anesthesia lasts several times longer than at night; retention of alcohol in the blood increases rapidly after about 10 P.M.

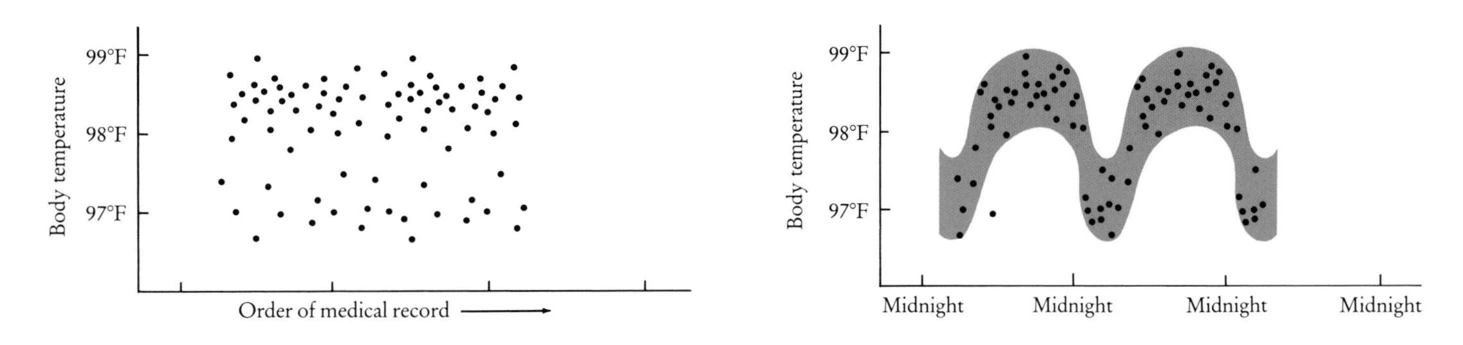

Normal limits for a diagnostically useful measure (body temperature) are plotted at left without regard to time of day, on the assumption that the norm and the clinically acceptable range around it are independent of circadian phase. But suppose the data points are plotted according to the time of day when each patient was sampled. Then circadian phase (or time of day, a good enough approximation) is thus taken into account. It is discovered that the norm varies rhythmically and that at each phase there is less variation around the norm in healthy individuals than appeared in the indiscriminately pooled data. An aberration then stands out more clearly.

in the day. Allergic responses are much more violent and are triggered much more easily in the wee hours of the night than in afternoon. Your liver is able to keep blood levels of alcohol much lower in the early evening than in the morning, a matter of potential interest to jet-lagged evening social drinkers.

Basic diagnosis might be made easier by systematizing clinical norms on a rhythmic basis. Body temperature, for example, is not normally 98.6°F at night: a "normal" reading at 3:00 A.M. would signal a fever. Again, Addison's disease and Cushing's syndrome are aberrations of adrenal function (insufficiency and excess, respectively), so diagnosis entails measurement of plasma cortisol levels. It is now commonly understood that a cortisol measurement has little value without knowing the time of day when the blood sample was taken.

Not only diagnosis, but therapeutic measures as well can be more effective when circadian variations are taken into account. Because many kinds of growing cell have a preferred time of day for replication of their DNA, circadian variations can be even more conspicuous in the toxicity of diverse drugs and irradiations whose purpose is to kill dividing cancer cells. It is not unusual for 80 percent of a population of test organisms to survive a dose at one time of day that would be lethal to 80 percent at a different hour. These grim statistics are richly promising when one thinks of ways to

(continued on next page)

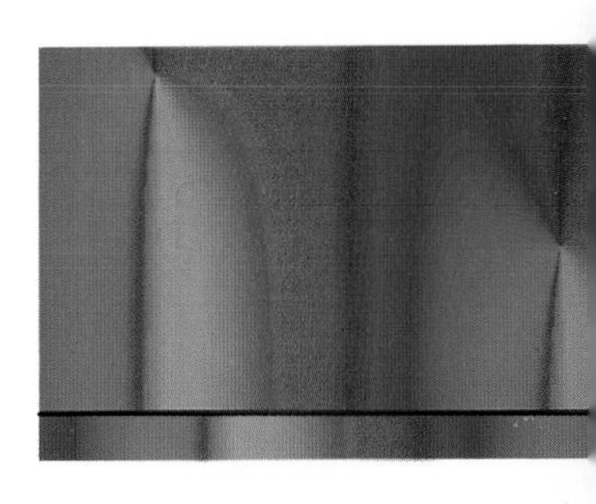

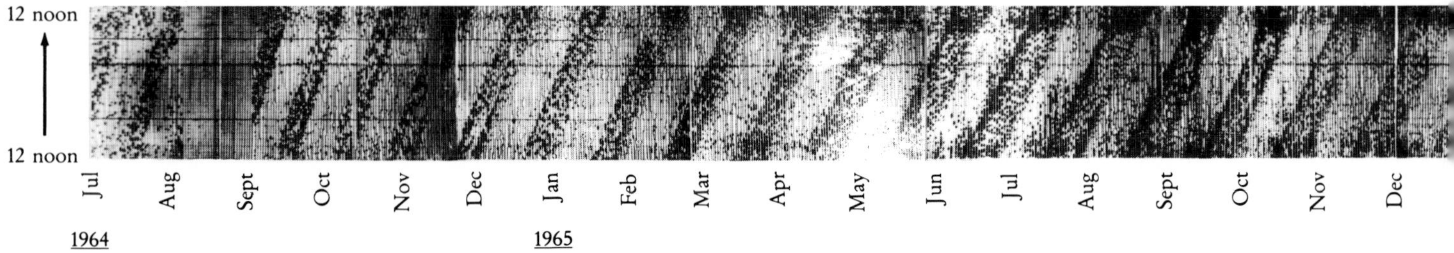

12 noon
12 noon
Jul Aug Sept Oct Nov Dec Jan Feb Mar Apr May Jun Jul Aug Sept Oct Nov Dec
1964 1965

selectively kill tumor cells without killing healthy cells. Edhard Haus and co-workers achieved a marked increase in the survival rate of cancerous mice by administering the same drug in the same daily total amounts, but concentrating the dose into a time of day when cancerous cells seem more susceptible than normal cells.[3]

Doctors and veterinarians have been aware for a long time of the importance of timing in hormone therapy. For adrenal deficiency, for example, it is routine practice to give human patients cortisone injections in the morning, when the adrenal is making its daily maximum effort. The morning schedule is also appropriate for dogs. If only a single dose must be given once a day, it should be in the evening for cats, whose circadian rhythms are differently organized. Given at other hours, the infusion of cortisone suppresses what little effort the deficient adrenal can muster, and the patient eventually succumbs to the chronic dependency called Addison's disease.

Rhythmic regularities connect brain hormones, sleep disorders, and more serious psychiatric problems, including clinical depression. They lead Thomas Wehr, Frederick Goodwin, and Norman Rosenthal of the National Institutes of Health in Bethesda, and Daniel Kripke of the San Diego Veterans' Administration Hospital, to suspect that these previously untreatable maladies are often secondary to disorders of circadian timing.[4] Some forms of depression already seem to respond to properly timed exposure to sunlight or its artificial equivalent.

"The trouble is," said the chief diagnostician, "we don't know what the trouble is.—it is impossible to tell whether the patient has clockitis, clockosis, clockoma, or clocktheria. We are also faced with the possibility that there may be no such diseases. The patient may have one of the minor clock ailments, if there are any, such as clockets, clockles, clocking cough, ticking pox, or clumps. We shall have to develop area men who will find out about such areas—"

JAMES THURBER, "THE THIRTEEN CLOCKS"

This illumination is believed to act through its effect on the secretion of melatonin, a brain hormone intimately involved with circadian timing. Light substantially brighter than common indoor artificial light is required for normal diurnal control of melatonin secretion. But contemporary Americans seldom expose themselves adequately, and then only at times of day that differ greatly from one person to the next.[5] It seems reasonable to wonder how common circadian disorders may be.

Mar Apr May Jun Jul Aug Sept Oct Nov Dec Jan Feb Mar Apr May Jun Jul Aug Sept
1967

Freedom and Lock-Step in Body Clocks

Alternations of sleep with activity maintained a 24.8-hour period in a blind squirrel monkey from September 1964 to September 1967.

The 25-hour clock that we share with other primates was revealed most clearly only 20 years ago, not in humans, but in a blind squirrel monkey living in Curt Richter's laboratories at the Johns Hopkins School of Medicine in Baltimore.[6] Still exposed to all diurnal influences while freely roaming the laboratory, but unaffected by light, the monkey alternated between sleep and wakefulness just as Richter did, but not at the same period. The blind monkey proved to have her own internal clock, which adhered to its unique period almost as faithfully as its human captors adhered to the day/night cycle of their rotating planet. The raster plot on this page records three years of the monkey's alternating activity and recumbence, with a period sometimes as long as 25 hours and 5 minutes, sometimes as short as 24 hours and 38 minutes. This range spans only two percent of the mean period, 24 hours and 46 minutes. Even in weeks when sleep was badly interrupted by construction projects in the lab, regardless of vacations, summer heat, and changes of overall health, the monkey's rhythm drifted later each day by 46 minutes on average, repeatedly scanning fully across the 24-hour width of the raster. For a few days each month, the monkey would seem to be on the same work shift with the doctors. But then the continuing gradual drift of its shift became apparent: she became a "night-owl" from the human viewpoint, just as they became "early-birds" from her viewpoint. Another two weeks, and the monkey's schedule had drifted so late that even her breakfast time occurred late at night: night-owl had metamorphosed to early-bird. Another week, and her habits were fleetingly in harmony with the doctors' again. This cycle repeated about once a month for years. The monkey kept her own internal time, not paced by the sunrise or the moonrise or any human schedules in the lab.

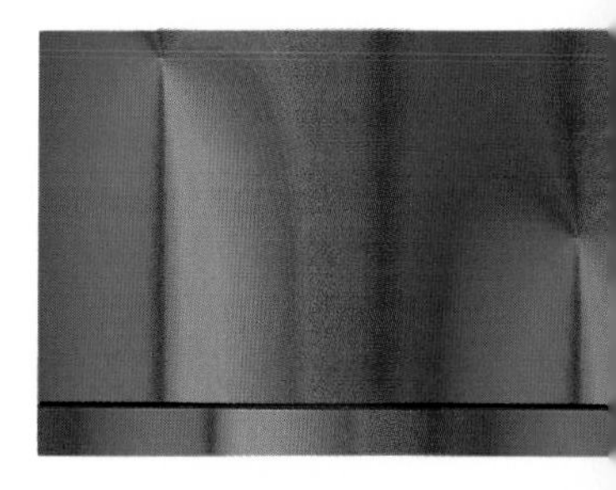

This indifference to external time turns up also in the body clocks of some blind humans. The raster plot of sleep and wakefulness on this page was recorded by a postgraduate student in biostatistics.[7] This young man had lost his vision at an early age. As a student he suffered for years in fruitless struggle to overcome chronic insomnia and daytime sleepiness. Curiously, this evil was upon him only for two or three weeks at a time, about once a month. He had kept a diary, which proved useful when he finally sought medical help. Raymond Miles and colleagues noted that the student's daytime naps drifted later by an hour every day until they merged with night sleep. A week later, they reappeared in the morning, and night sleep began to be riddled with insomnia again. The cycle repeated every 25 days or so.

Henry David Thoreau had already diagnosed the case in 1854: "If a man does not keep pace with his companions, perhaps it is because he hears a different drummer. Let him step to the music which he hears, however measured or far away." Suspecting some 25-hour drummer, the consulting doctors invited the young man to desist from his struggle to keep pace with society's 24-hour rhythm. They suggested sleeping only when he felt sleepy, while protected from interruptions in a convenient hospital. His habits immediately became regular for the first time in many years: full alertness throughout the day, solid sleep at night. But "day" and "night" no longer meshed with day and night outdoors; the patient's regularity had a 25-hour rhythm.

Returned to civil time for a month, his sleep was again fragmented despite heroic efforts with caffeine and sleeping pills. Set in context now, his daytime naps can be seen as the camouflaged steady continuation of an internal 25-hour cycle. Something like this syndrome may afflict about half of all blind individuals.

Not only blind people have 25-hour internal clocks: that much is the same in perfectly normal individuals. Though the internal clock's natural pace is normally lethargic by about an hour a day, it is normally sensitive to the 24-hour day/night cycle and thus keeps pace with it. On the next page you can see this happening for the first time in a newborn baby girl's gradual consolidation of a 25-hour rhythm, and its subsequent adaptation to the 24-hour rhythm of her environment. However, many blind individuals, and some others as well, lack this capacity to reset their internal clocks by an hour each day. Some otherwise normal adults also fail to synchronize to the earth's rotation and the society of their companions. The sleep raster on page 32 was painted by a research analyst with normal vision who simply could not keep up

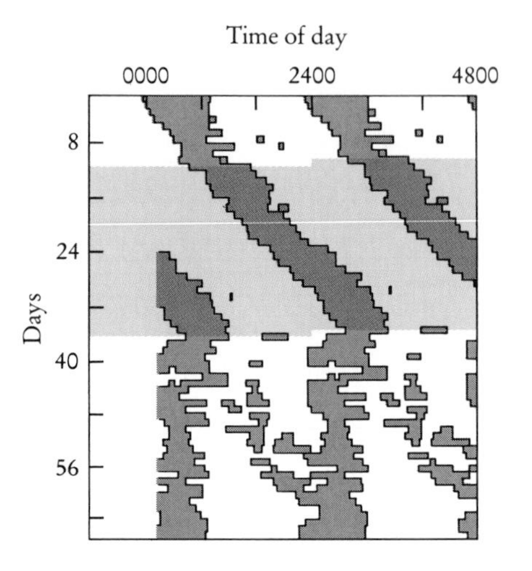

A blind subject slept during the shaded areas of this double raster plot. During untinted intervals, when the subject lived in his usual environment of home, work, drugs, and alarm clocks, sleep was fragmented and work time was interspersed with naps. During the interval shown by a color tint, he made no attempt to maintain a 24-hour cycle; his innate 24.8-hour cycle stands out clearly, and extrapolates plainly into the intervals when he did attempt to maintain a 24-hour cycle.

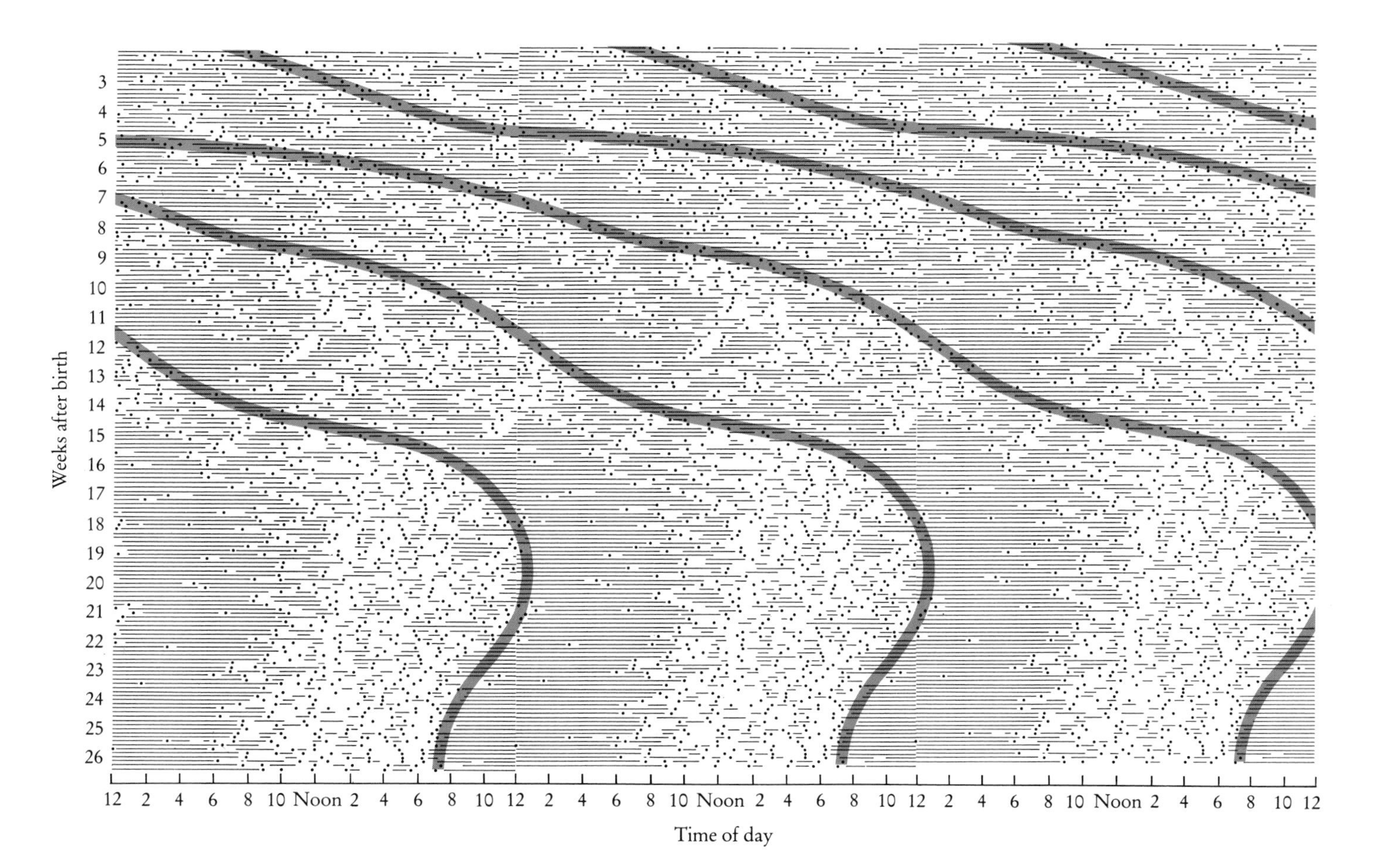

Sleep/wake cycles of a suckling infant during the first half year of life gradually emerge and lock onto the 24-hour pattern of its new surroundings. This was a first baby and its parents were sufficiently indulgent to permit it to set its own schedules of sleep and feeding. Dark lines are sleep, dots are feedings.

with the conventional 24-hour cycle.[8] He drifted later every day for years, though at a lesser pace during the days when he was temporarily almost in step with everyone else. His average period here is about 25 hours. (Curiously, about two months after this study ended, he gradually became able to synchronize with his surroundings. What made the difference? No one knows, though it happened about the time that he switched to a low-glucose diet.)

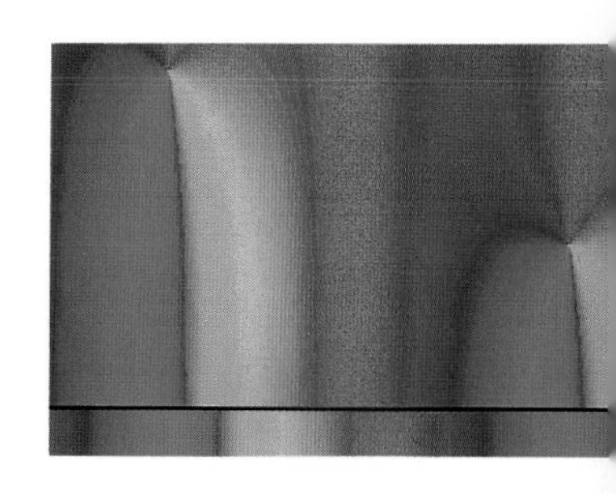

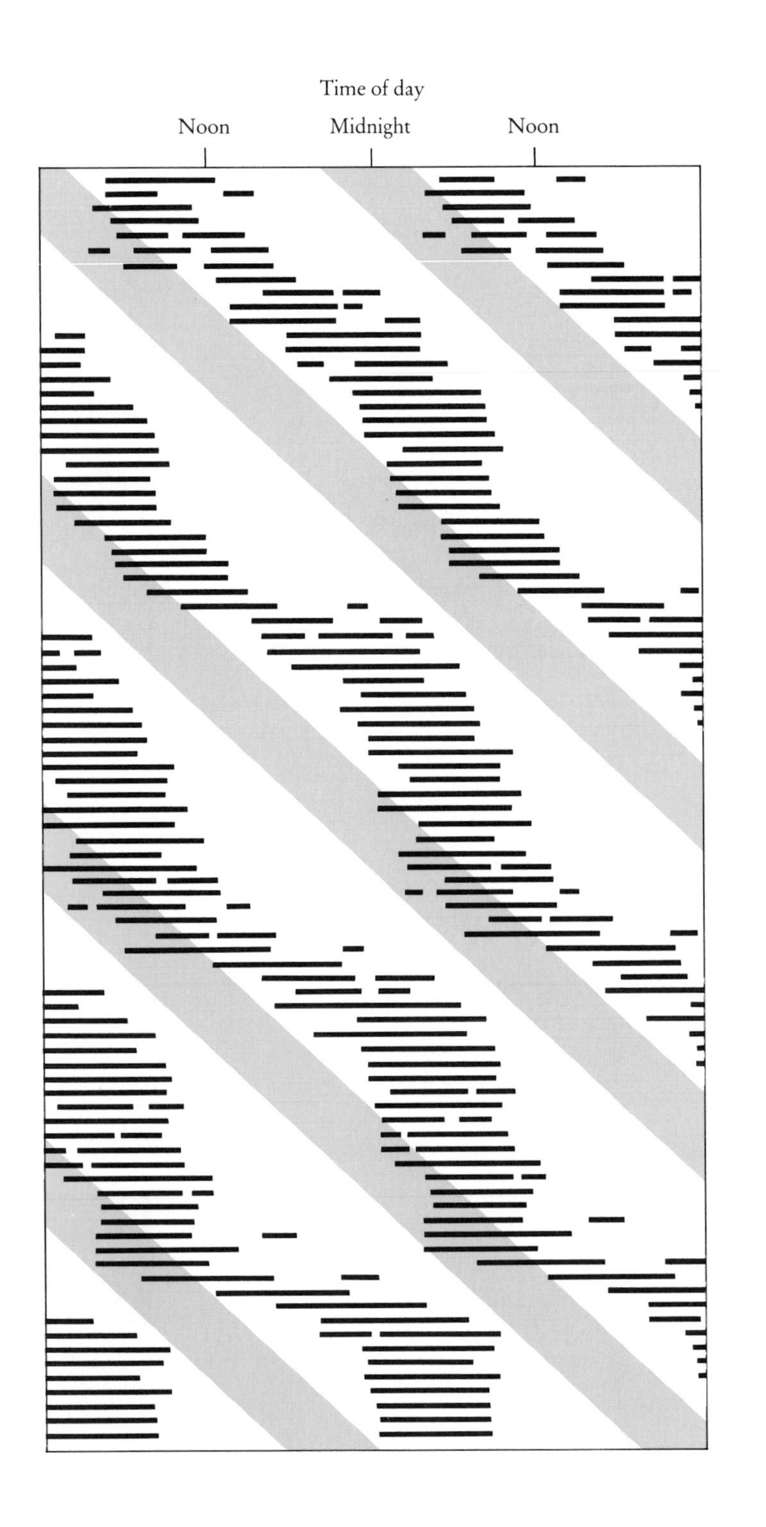

The double-plotted sleep raster of a normally sighted person whose period always exceeds 24 hours. During days of near-synchrony with conventional sleep time, his period approaches 24 hours, but not enough. His sleep time still drifts later slowly, then rapidly later while it is opposite to normal. He almost never awakens during the shaded interval in his 25-hour cycle.

Another such unsynchronized man, with a mean period near 26.5 hours, enjoyed the companionship of a 24-hour woman—except when they were at opposite phases of their circadian cycle.[9] The raster on the next page shows an interval in their struggle to stay synchronous at a compromise period near 24.7 hours. It worked most of the time, but at least once—perhaps once in each month—they lost it: she gained half a day while he lost half a day, so that they rejoined by quickly getting to be a whole day apart. Over the whole month, then, each was still dominated by the period of a steadfast clock: his apparently internal, with an abnormally long period of 26.5 hours; and hers staidly external, with the period of the day/night cycle. The difference was only 6 percent, but irreconcilable.

In normally sighted people exposed to a 24-hour cycle of alternating light and darkness, such cases are exceptional. But when someone deliberately retires from the merry-go-round of the earth's surface, for example, to the timeless quiet of an underground cave or even just to a windowless room, the rhythm of sleep and wakefulness almost always reverts to its natural pace, about 25 hours. It seems that heaven and society are always hurrying us along: to keep up with the 24-hour world we must hasten, leaping ahead of ourselves by an hour every day to keep up. Much as the Red Queen remarked through Lewis Carroll in 1871, "Now, here, you see, it takes all the running you can do to keep in the same place" on this planet that spins faster than we do (and it used to spin even faster in the geologic past, before genus *Homo* emerged). Some people's inner clocks give up the race in favor of their own integrity. How and why, we don't yet know.

The reversion to the internal standard of time is less obvious in humans than in some other mammals, perhaps because our conscious behavior and our physiological processes are more independent of one another. A subtle test is sometimes required to disentangle the robust biological tide in humans from the many other factors that affect sleep timing or any other single rhythm. The strong 25-hour component of regularity in the ups and downs of body temperature, for example, can be reliably extrapolated months ahead, but at any given hour, our tendency toward the predicted temperature may be overwhelmed by the transient effects of exercise, showering, or eating ice cream. Our habits of sleep and waking are also subject to direct conscious control, and their irregularities reflect the vicissitudes of the day. Nonetheless, our internal circadian clock vigorously dominates the long-term trends of our own sleep, and guarantees that we will sleep (or desperately wish to)

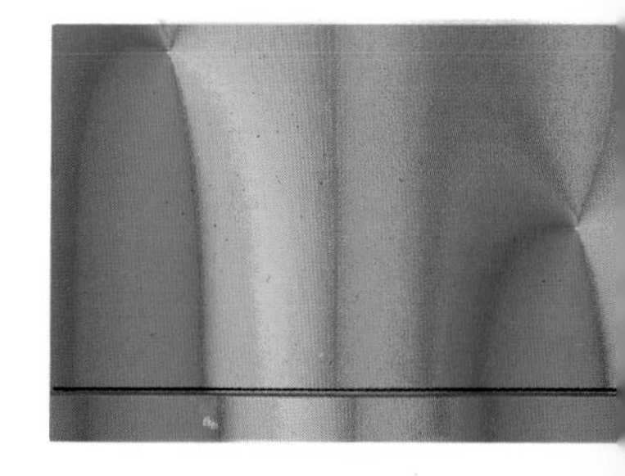

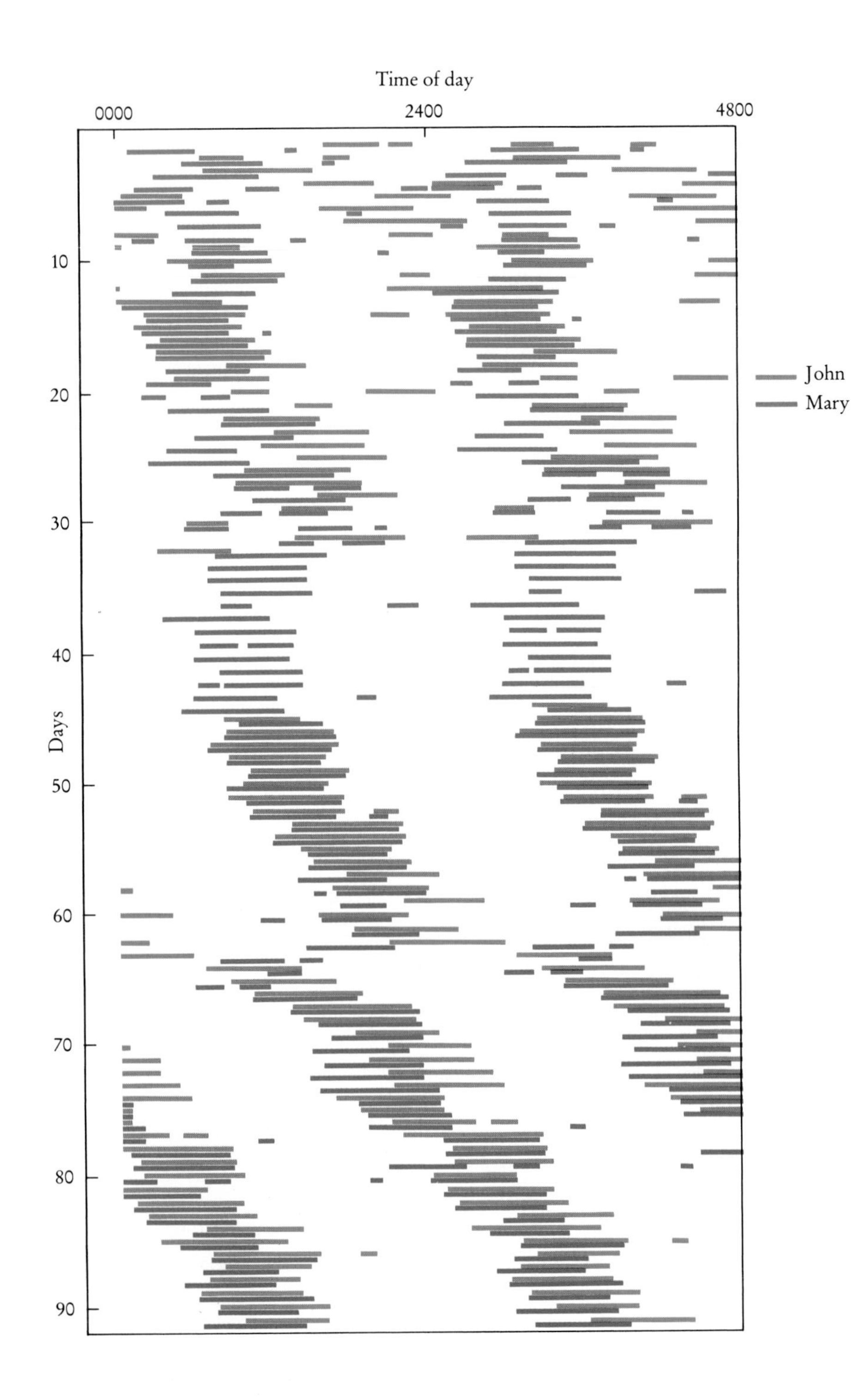

Three months of double-plotted sleep raster showing in blue Mary's attempt to depart from the customary 24-hour cycle in order to stay synchronous with John's 26.5-hour cycle (in red). From day 32 to day 45, she lived alone while John visited his parents. During other times they compromised near 24.7 hours, but at about day 63 John involuntarily delayed half a cycle while Mary advanced half a cycle, so each kept his or her own period on average.

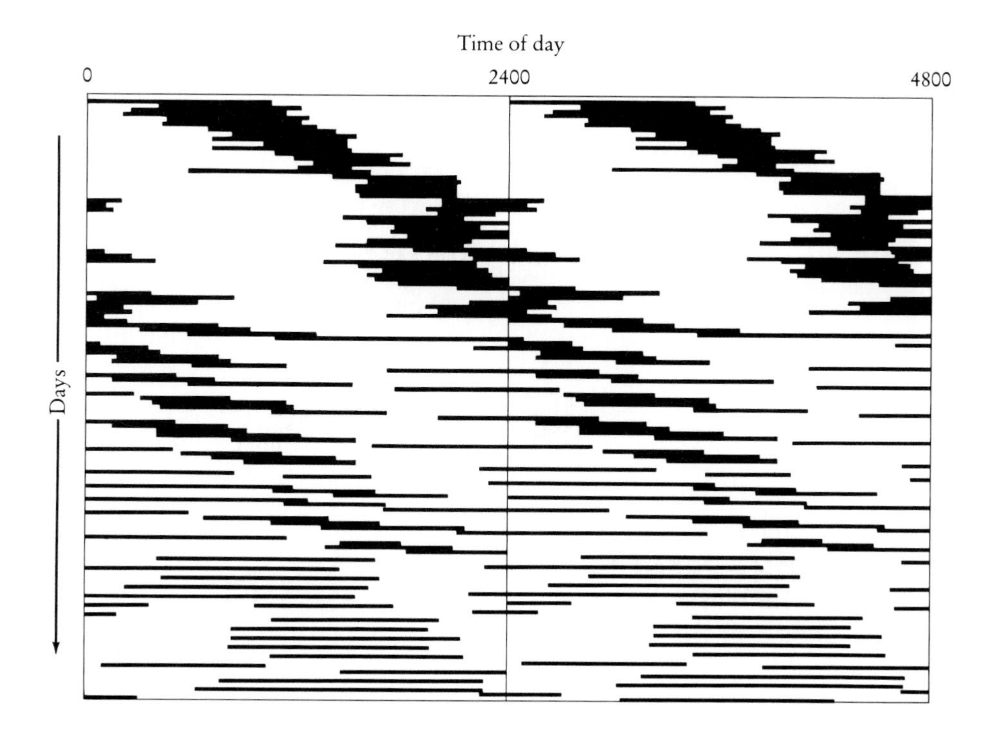

Double-plotted sleep raster of subject J. Chabert during a cave experiment seems to defy simple interpretation. "Time of day" starts from an arbitrary zero.

about once in every turn of the earth on its axis. What is perhaps even more remarkable, once sleep is achieved, the internal clock is preset for spontaneous waking at a phase in the circadian cycle that depends on the phase when we went to sleep.

Automatic Setting of the Inner Alarm Clock

In early January of 1969, 23-year-old Jacques Chabert undertook to live for half a year beyond the reach of human timekeeping, in the most physically timeless environment anyone could think of: a cave constantly at 6°C and 100 percent relative humidity, 65 meters beneath the roots of grass and trees near Nice and Cannes, in southern France.[10] Looking at his sleep raster, anyone's dominant impression would be arrhythmia: the regularity of sleep/wake timing has apparently broken down in the absence of the usual 24-hour driving influence. Without that external clock to assign a regular bedtime and a regular time for awakening, sleep and wake durations appear to vary at random. Do they really or do we only lack the deciphering key?

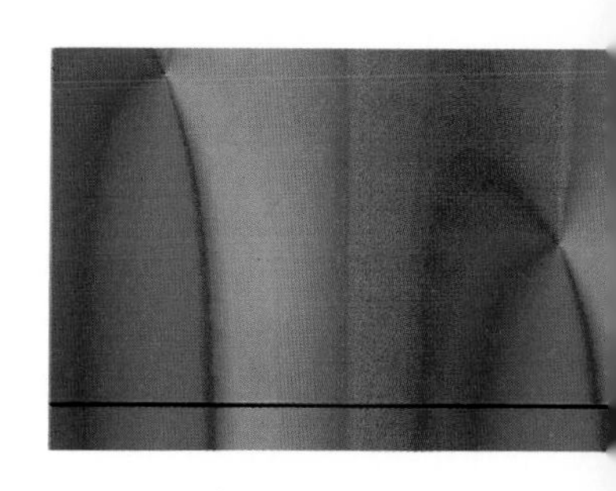

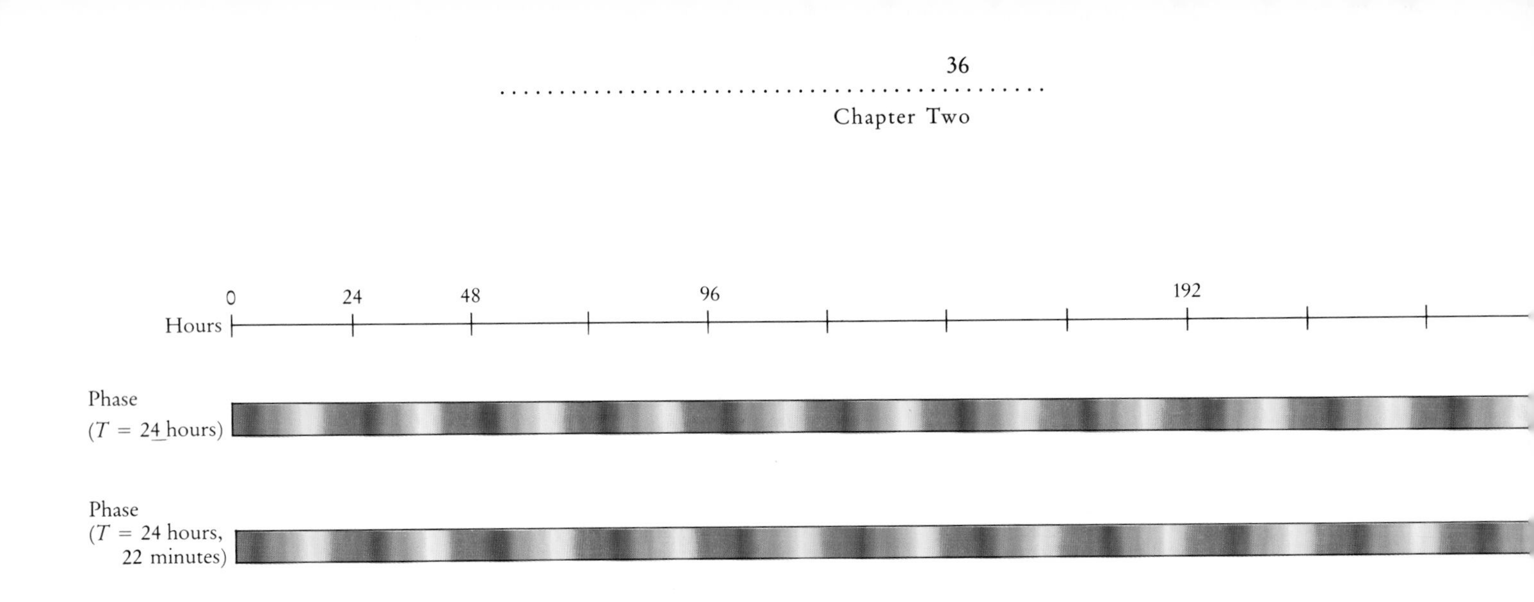

Let us suppose that the man has a circadian clock. Whatever its unknown period, it relentlessly ticks off the hours while our resolute subject has chills and fevers, darkness and lantern light, exultations and depressions, good novels and dull ones, and chooses to sleep when he will. If the cycling of this circadian clock plays some role in timing his involuntary awakening, then his sleep duration might depend systematically on the point within the cycle (the phase) when sleep began.

We need first to assign a phase to every hour when sleep begins or ends throughout the whole 127 days of temporal isolation. Unfortunately, the phase itself is not directly observable, except to the weak extent that it influences body temperature. But one thing about this internal master clock is very reliably known: it advances with remarkable steadiness. If we are brave enough to assume absolute constancy of the clock's period T, we can adopt a *phase ruler* of length T, and, placing copies end-to-end along the whole 3000-hour caving record, we use it to measure off time in units of that standard cycle, thus assigning a phase within the cycle to each hour. Phase rulers of slightly different lengths will assign similar phases to hours near the beginning, but will assign substantially different phases to hours far to the right. We don't know the correct period exactly, but we can find out whether there *is* one as follows.

Note the phase when each sleep began. Plot each interval of sleep as a dot, positioned horizontally by that phase and positioned vertically according to the subsequent duration of sleep until spontaneous waking. This shows the putative dependence of sleep duration on the phase of sleep onset, for the assumed period T. If the circadian clock plays no major role in sleep timing or if we guessed its period wrong or if it does

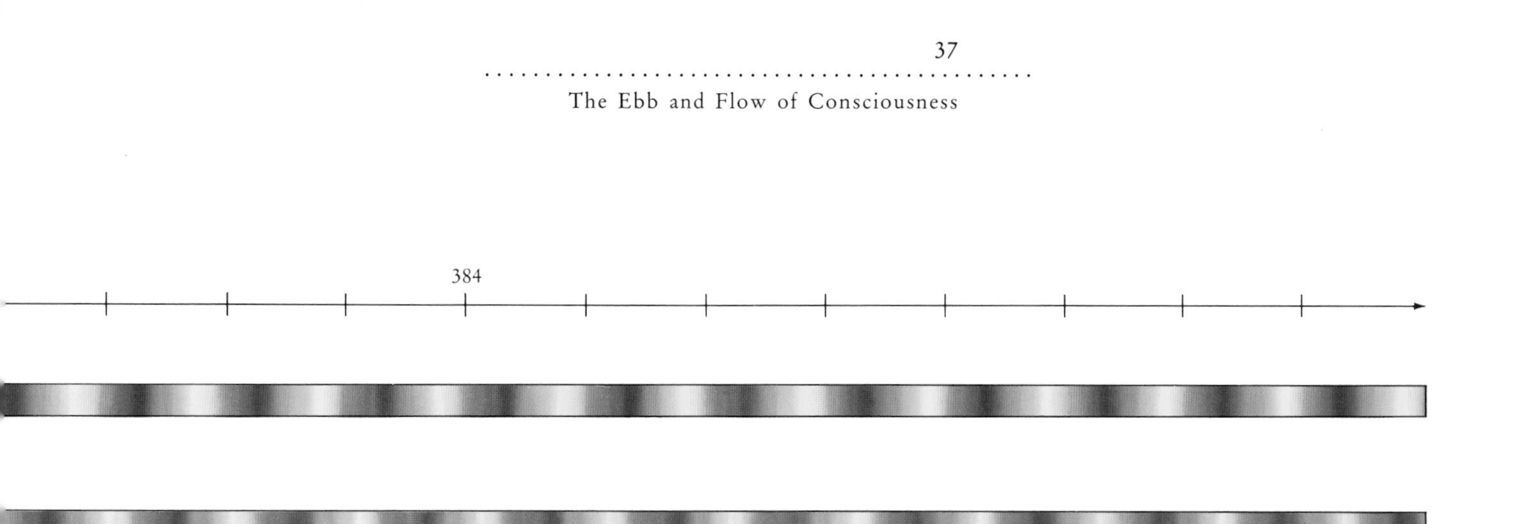

not in fact cycle inexorably with such absolute precision, then the cloud of dots will resemble nothing more than a shotgun blast ranging horizontally across the assumed cycle duration, and ranging vertically from the shortest sleep (about 3 hours) to the longest (about 18 hours). And that is just what the plots all resemble, for guessed periods around 24 hours (2000 minutes)—except within about 10 minutes either side of $T = 1458$ minutes = 24 hours and 18 minutes. In that narrow range alone, the data points suddenly fall into a pattern: the duration of sleep is fairly predictable, given the time of sleep onset, if and only if we use the proper period to establish the phase of the internal clock. This predictability reveals the rhythmic organizer hidden in the subtle pattern of unsynchronized human sleep and waking. The period is fairly sharply defined; if it gradually drifted as much as 10 minutes during the 127 days, then the phases assigned at the end of 127 days would have been altered by about a half cycle, so the data would have been badly scrambled.[11] Evidently, the postulated clock is more constant than that, even comparable in precision to the inner clock of Richter's monkey, which, over the course of two or three years seldom strayed as much as 10 minutes from the 24.8-hour mean period.

By laying phase rulers end to end, a phase is assigned to each hour in a long continuous record (top line). Starting at time 0 with a red phase (arbitrarily), red repeats at 24-hour intervals (middle line). If phases are reassigned using a slightly longer phase ruler (period 24 hours and 22 minutes, bottom line), a substantial difference accumulates toward the right.

Some dependency of sleep duration on the timing of sleep onset would be expected if, for example, we tend to wake up at a standard phase in the circadian cycle, or if we tend to sleep longer after retiring late and wake earlier after retiring early. In fact, however, neither of these dependencies is reliably borne out by anyone's data. The usual notion that sleep serves a restorative function in proportion to its duration cannot be supported from measurements of sleep as prolonged as we commonly enjoy at night.

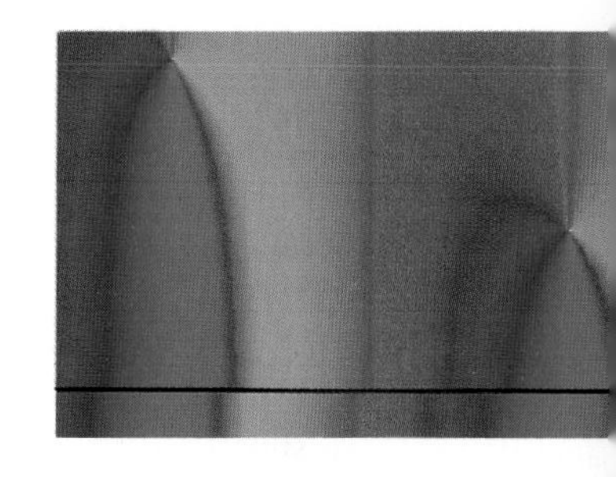

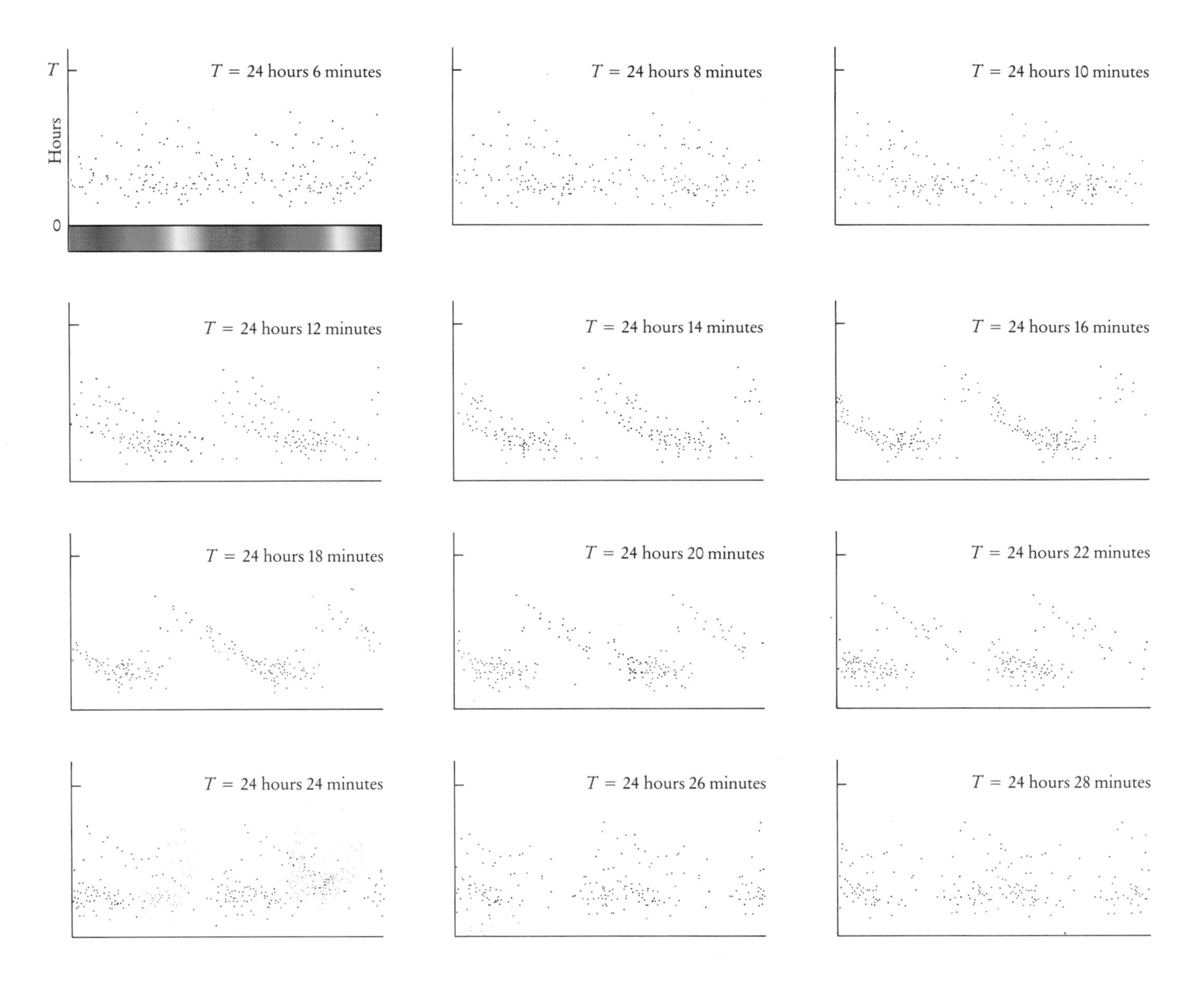

In each panel, the data of the raster on page 35 are replotted to show sleep durations (on a vertical scale from 0 to T hours) for sleeps begun at each phase (represented horizontally on a phase ruler of colors, double-plotted) of a postulated cycle of period T. The assumed period T increases by 2 minutes from one panel to the next. Near T = 24 hours and 18 minutes, a functional relationship emerges, revealing the presence of a real cycle and its effect on the timing of sleep.

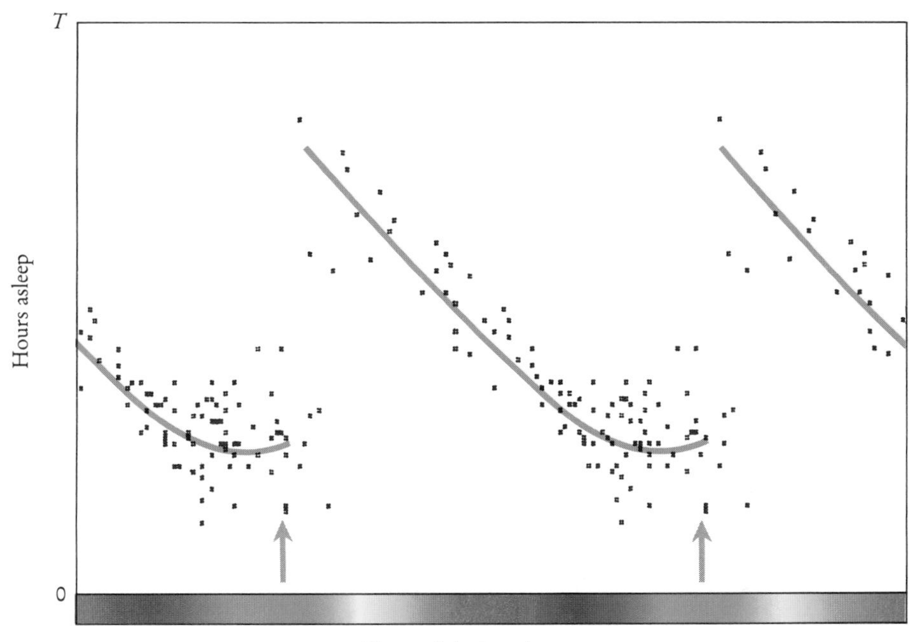

Panel T = 24 hours and 18 minutes, with a scoop-shaped mean curve drawn through the data. At the top of the discontinuity, there are apparently two choices: sleep very briefly or very long, but in either case do not awaken after a middling duration of sleep.

The surprising result of the caving experiments is not that there is a pattern, but that the pattern actually found is not one of those expected, and that the scheme for detecting it reveals an inner clock of remarkable precision. This dependency was first observed somewhat obscurely in a study of the irregular sleep of train drivers: the duration of naps or attempted full sleeps varied systematically with the time of day.[12] Several years later, Charles Czeisler and Elliot Weitzman began to analyze caving records such as Chabert's by simply *assuming* an underlying internal clock, then trying to guess its period to minimize the variance of sleep durations at each phase.[13] The magic period that emerges, in the range of 24 to 25 hours, is the same as the period of the body-temperature rhythm. About 1980, Jurgen Zulley and Rutger Wever detected this pattern by reanalyzing Wever and Aschoff's records of temporal-isolation experiments.[14] Zulley plotted sleep duration against the timing of sleep onset in each subject's 25-hour rhythm of body temperature, rather than against a hypothetical phase ruler; he got about the same picture as in the graph on this page.

These temporal-isolation records, organized by the internal 25-hour clock, are only a variation on what you already know from casual expe-

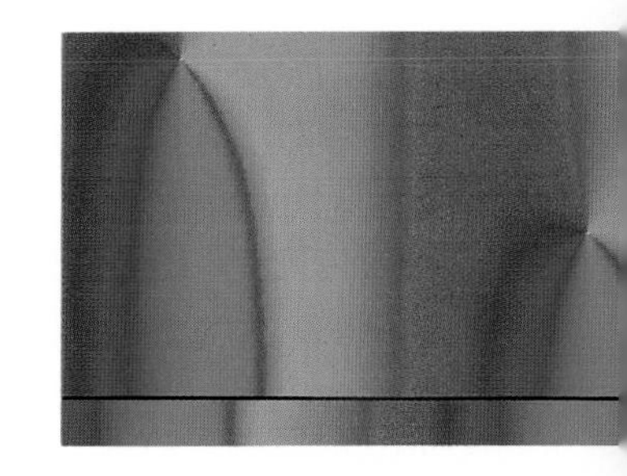

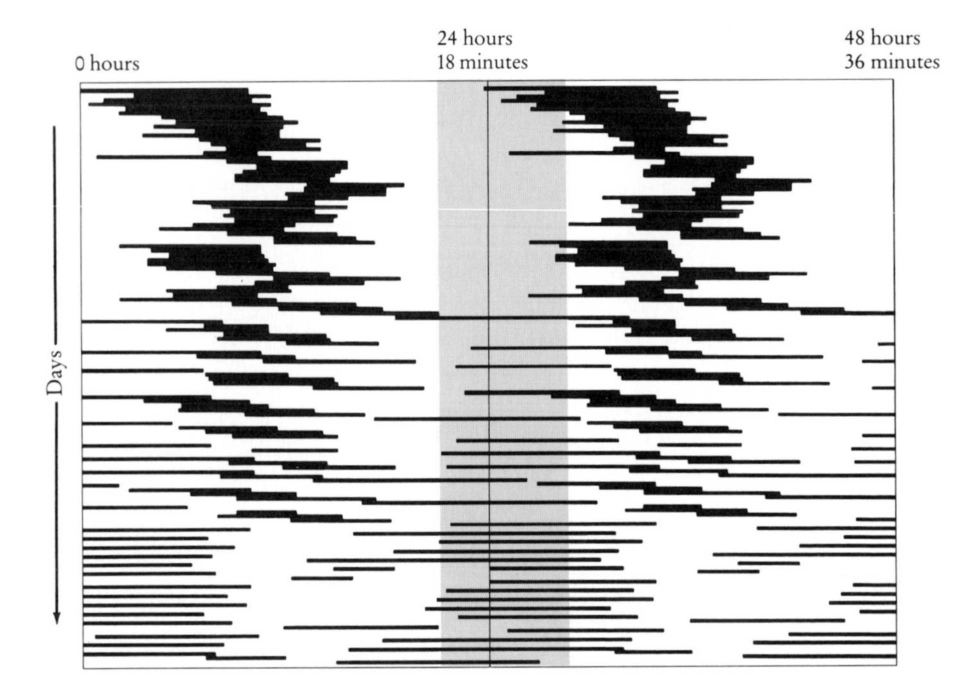

J. Chabert's sleep raster repeated from page 35, but tuned to a period of 24.3 hours rather than to the usual 24.0 hours. The shaded band is a range of phases within which spontaneous awakening (represented by the right end point of each black sleep bar) is rare.

rience under the discipline of a 24-hour cycle. During temporal isolation, much as in synchronized life outside, the earlier you go to sleep (up to a point), the earlier you will spontaneously awaken, though your sleep will also be slightly longer. However, if you retire too early, sleep duration jumps abruptly back to a nap as short as about 4 hours. This jump from longest to shortest sleeps shows that sleepers are unlikely to awaken spontaneously in the middle of the interval jumped over. A sleep that starts at this critical phase in the cycle may terminate 5 hours later or 17 hours later, but not 11 hours later. This forbidden interval for spontaneous waking appears on a raster plot as a range of phases almost devoid of sleep-endings. Burglars take note!

Although the emergent pattern—a ski slope punctuated by a discontinuity—was not among those foreseen, it might have been. This pattern is common in physiology wherever the timing of a discrete event, such as transition from sleep to waking, is governed by some process encountering a threshold condition, if that threshold is subject to some rhythmical influence. The central idea can be described by a simple metaphor.

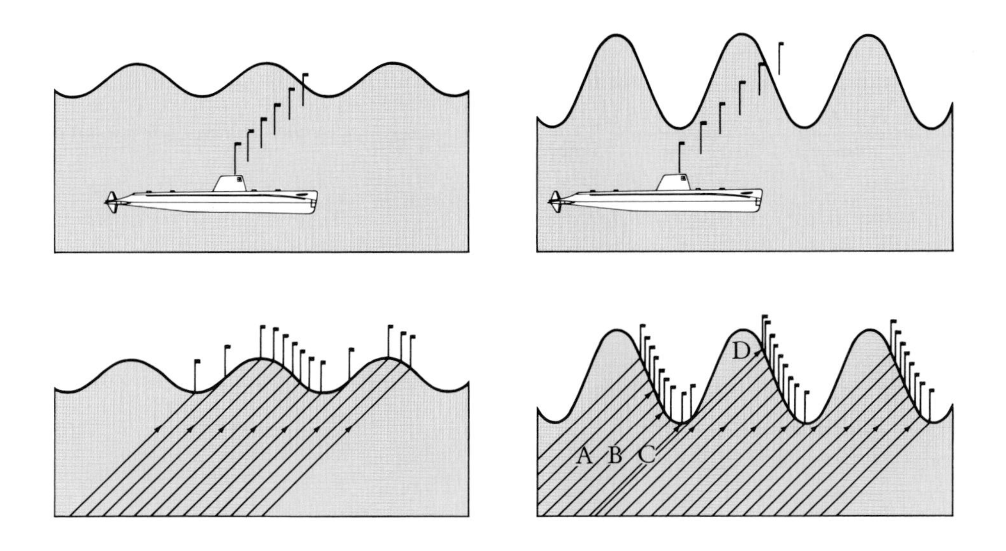

A submarine ascends toward the rhythmical interface between water and air. If the waves are shallow, ascents vary only slightly in duration, and penetrations occur at all phases of a wave, only slightly biased toward emergence on the descending side. If the waves are big enough so the water level sometimes rises faster than the submarine, then durations of ascent vary more and penetrations occur mostly while the water level is falling.

How a Smooth Rhythm Can Trigger a Discrete Event

A cruising submarine slowly ascends from the sea floor, periscope up but still well beneath the long rolling waves. When will it first break the surface? If the ascent is nearly vertical or the waves are quite shallow, then the breakthrough may happen anywhere on a wave. But if the ascending slope is shallow or the wave troughs are deep enough, then breakthrough is sure not to occur on the steepest rising slope. So what?

The diagram above represents a snapshot with the camera moving along with the waves, so the waves seem stationary. The submarine (periscope shown in multiple exposures) is moving forward as it rises relative to the stationary waves. Its periscope first breaks through to air somewhere on the wave surface. Just where depends on timing. If the surface slopes much less than the periscope's path of ascent as in the leftside drawings, then, in a hundred trial ascents, the breakthroughs will be only slightly more common where the waterline slopes gently downward than where it slopes upward.

But if the waves are steeper—steep enough so that the water rises faster than the submarine ascends as in the rightside drawings—then the rising portion of the wave is immune to penetration from below. A hundred trial ascents will break the surface only in bunches, completely missing the interval when the wave surface slopes up more steeply than the periscope's ascent.

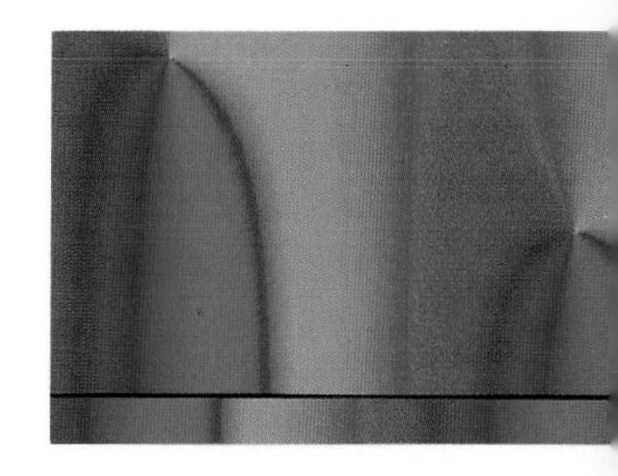

The width of one full wave from crest to crest represents the duration of one cycle of the circadian clock. The wave represents the rhythm of something in the human brain—no one yet knows what—that, like body temperature and a dozen other measured physiological variables, smoothly rises and falls once in every cycle. This "something" sets a threshold at which the sleeper (the rising submarine) awakens. The various parts of the wave represent the changing threshold level at different phases in the circadian cycle. What we surmise from this metaphor is that there may be forbidden phases, phases when the periscope can never break through. Such forbidden phases were observed in temporal-isolation data. What else about those data might be reflected in this picture of breakthrough timed by a periodic influence?

Notice the final altitude at the moment of breakthrough: this shows the total duration of ascent, given a fixed slope. In the illustration on page 41, ascent B starts somewhat later than ascent A and will hit threshold at lower altitude, that is, sooner. Ascent C, still later, will hit still sooner. But ascent D is *too* late: by the time the submarine approaches the moving surface, the surface is rising again, and too fast. Then the trough has been missed and ascent must continue through deep water to break through after the *next* peak. The duration of ascent is least for C, who encountered a trough just before the steeply rising side of a wave; it is longest for those, like D, who missed the trough and are already impatiently waiting at high altitude when the surface first descends to meet them. A continuous series of 100 submarines starting their ascents later and later will break through to the air not in continuous succession, but in bunches spaced apart by the period of the waves. This arrangement works as though the waves open and close a gate—but only if the waves are steep enough relative to the submarine's ascent. A weak rhythmic influence only modulates the breakthroughs without bunching them discretely. There is a critical steepness above which part of the cycle becomes strictly forbidden, as though an impenetrable gate were closed.

Men and women seem to time their spontaneous wakings by some such principle.[15] Our consciousness ebbs and flows in a daily cycle. Arousal from sleep in that cycle seems to be triggered when something, gradually changing during sleep, reaches a threshold level that varies, as does almost everything inside us, with circadian period. If that change begins at a given phase (when you begin to sleep), then it will reach threshold at a predictable later phase and (supposing no raucous alarm goes off first) you will then wake up. Your duration of sleep may vary

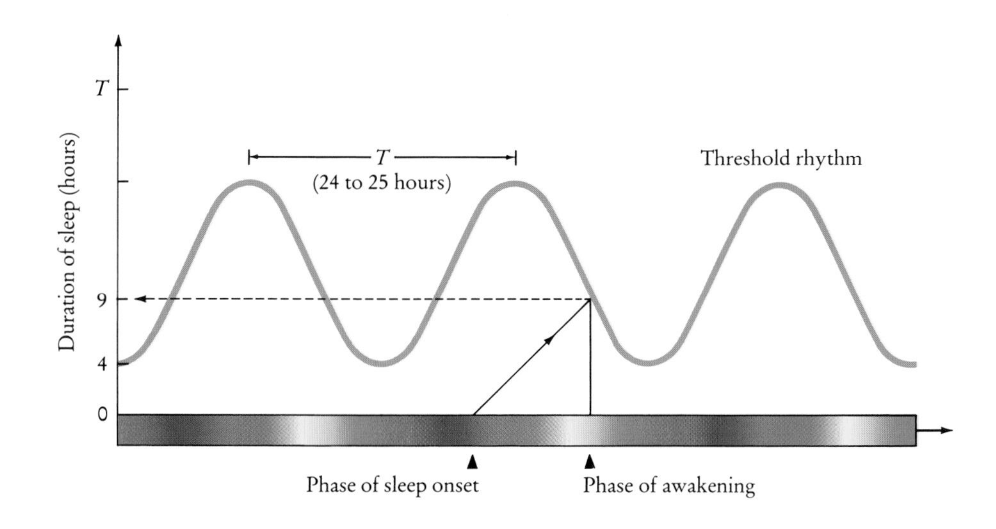

Sleep may be imagined to terminate when some quantity (in a broad sense, *restedness*) that increases during sleep reaches a threshold. Since most things vary diurnally, so might this threshold. In this diagram, some sleeps could be as long as 18 hours, some as short as 4 hours, depending on when the sleep begins. A 9-hour sleep is shown.

quite a lot according to the timing of sleep onset; the average is 8 hours only because people usually go to sleep around the corresponding phase of their circadian cycle. If your circadian variations are strong, then your sleep may be as brief as 4 hours or as prolonged as 18 hours, depending on when in the cycle you chose to start sleep. You can switch from the shortest to the longest sleeps by delaying the start of your sleep just enough to miss the trough in the threshold rhythm. Consequently there is a range several hours wide in the circadian cycle when you are almost sure not to wake up spontaneously. If for some reason your physiology is less phase-dependent (your threshold rhythms are shallower) then your sleep duration may not be as conspicuously affected by circadian time as by other things (tiredness, for example). In that case, sleep duration would decrease gently as sleep onset begins later, and then with still later sleep onsets, it would gently increase back again: no discontinuities, no zone of forbidden waking. This might be your condition after long exposure to continuous daylight in the arctic summer (which does result in peculiarly attenuated rhythms in Eskimos living outdoors), or during jet lag (which temporarily confuses and attenuates normal circadian variations).

The timing of sleep termination, the onset of waking, is only one face of a coin. What about the timing of wake termination, the onset of sleep? Circadian principles might be inscribed on that obverse side similarly, but not identically: sleep timing and wake timing differ funda-

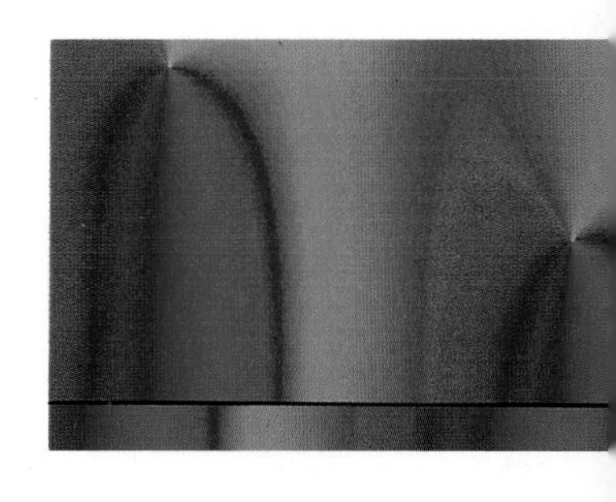

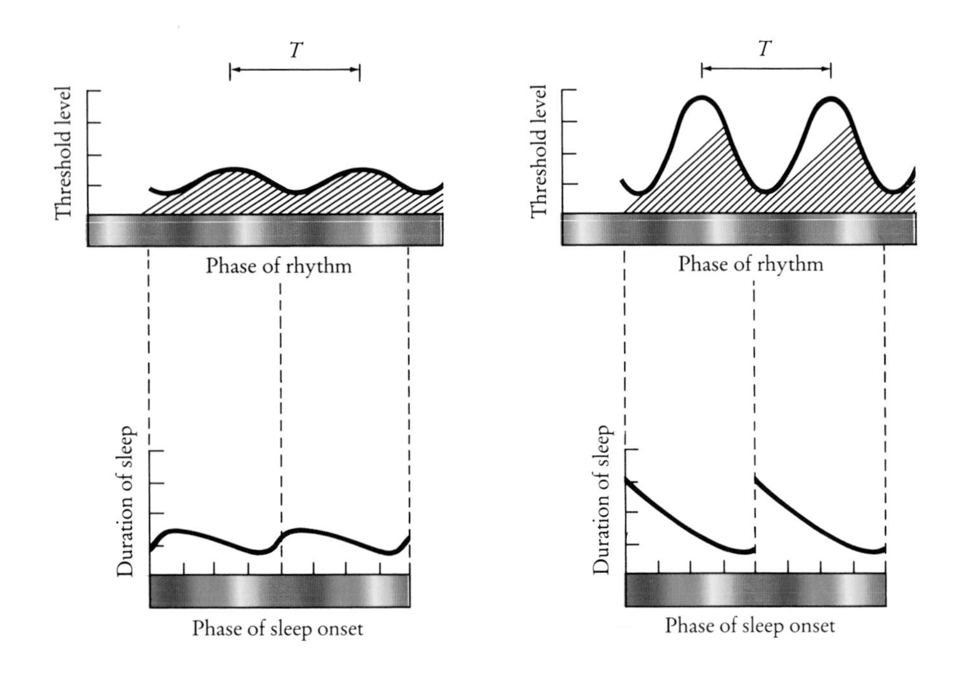

Top right: Time of awakening depends on when restedness (diagonal lines) first encounters threshold. If the threshold rhythm is strong enough, sleeps do not end during certain phases. *Bottom right*: The discontinuous ski-slope pattern of sleep durations.
Top left: If the threshold rhythm is weak, sleep may end at any phase. *Bottom left*: In that case, the dependence of sleep duration on the phase of sleep onset is smooth. This might be expected of people whose circadian rhythms fail to dominate their sleep timing—for example, people suffering from jet lag or chronic exposure to bright lights.

mentally.[16] There is a single wide zone of the circadian cycle when spontaneous waking is practically forbidden, but only a narrow range—actually two such, about 12 hours apart—when sleep onset, although not really forbidden, is relatively rare. Additionally, a plot analogous to the graph on page 39, but relating the phase of wake onset to duration of the next waking period, looks like a snowstorm. Not only is data variance enormous, but worse: no curve could possibly be drawn through the data because there are few data, if any, in the forbidden range of wake-up times.

Only a backward dependency has been found: the duration of waking prior to a given sleep onset is longer by roughly an hour for each hour that sleep onset is later. Beyond a certain phase of sleep onset, prior wake duration jumps discontinuously as the phase of prior wake onset avoids the forbidden range. Thus a backward ski-jump curve is outlined for the phase at *prior* awakening. But the data are scattered twice as broadly around it as they were for the original and predictive ski-jump curve of phase at *next* awakening. Sleep does tend to begin around the low point of body temperature, but only this vague statistical regularity connects it to the circadian cycle. Perhaps in humans voluntary control is exercised more vigorously at the end of waking than at the end of sleep, so that timing depends on many more factors than just the circadian clock.

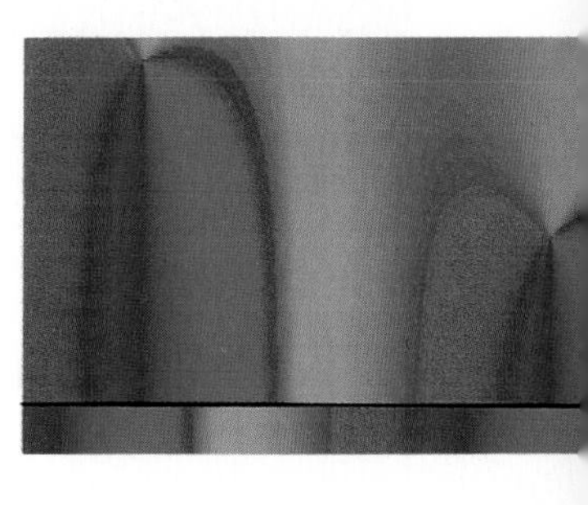

Gonyaulax cells near peak time in their diurnal glow rhythm make halos around the pilings of the old pier at Scripps Institution of Oceanography in La Jolla.

Chapter Three

Jumping Between Time Zones

Men occasionally stumble over the truth, but most of them pick themselves up and hurry off as if nothing had happened.

ATTRIBUTED TO WINSTON CHURCHILL

A clock is not much good if you can't pull out its stem and reset it. An uncorrectable biological clock would be almost useless unless the match between its period and that of the earth's rotations were perfect. Still, even if the clock were immune to disturbance by fevers or cold or hot weather and remained unperturbed during intervals of emotional and hormonal upset, there would inevitably be some small discrepancy between the internal and external periods. Suppose the discrepancy were as little as one minute per day. After 2 years, you would no longer be a diurnally active animal, but would be nocturnal, waking at dusk and going back to sleep at dawn. After another 2 years, you would be back in step with the world around you; another 2, and you would be nocturnal again. If your clock were 10 times more accurate, this 4-year cycle would become a 40-year cycle but you would still be nocturnal half the time. The only way you could stay in step with your time zone would be to become a global nomad, changing time zones just fast enough to circle the globe every 4 (or 40) years. Clearly, to maintain synchrony requires more than a close match of periods; it requires a cadence caller, a mechanism for daily cueing and resetting. The essence of any biological clock's utility is this phase resetting, its capacity to leap on cue from one time zone to another and so to stay in the right time zone without physical travel, despite the inevitable mismatch of periods.

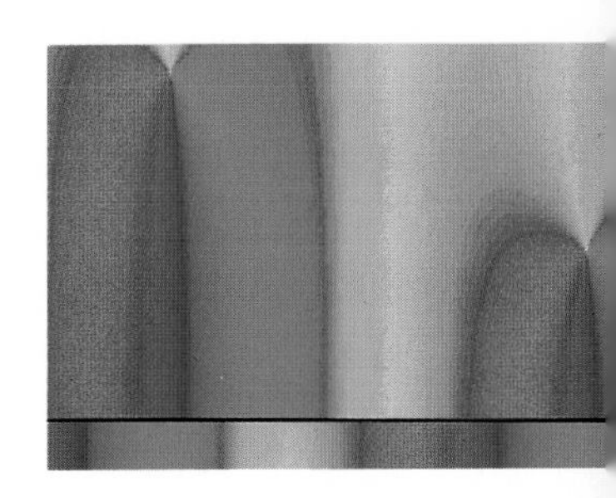

This process is called *entrainment*. One cycle can be entrained to the exact period of another by means of regularly repeated cues, much as

Flowering in the morning glory, as in many other plants, is timed by a circadian clock. In this four-hour sequence a single mature flower greets its morning.

dancers are entrained by the beat of the music. The flowering and fruiting cycle of many trees is entrained by the cycle of seasons. So is the growing and shedding of antlers by reindeer. For circadian rhythms, the usual natural cue is the daily interval of light. Exposure to light sets the internal clock back a little or nudges it forward. There are other ways to reset a biological clock, but the vast majority of living species have adopted the daily event that recurs most reliably—the alternation of light and darkness—as the event to which their circadian clocks respond most sensitively.

This generalization, that the cueing agent is light, has been questioned only for humans, partly because humans, like most other organisms, can and sometimes do take their cues from other senses in the absence of a dominant influence from light. But it was questioned mostly because the indoor "daylight" used in the first temporal isolation laboratories seemed not to dominate human circadian rhythms as strongly as light of that intensity does in other animals. Does that observation reveal that the human clock is uniquely indifferent to light? No; more likely it indicates only that the indoor light was too dim for the human clock to see it as daylight.

There is good reason to believe that the dominant circadian clock in humans is in the brain and is connected directly to the eyes and to the pineal gland. Alfred Lewy and collaborators at the National Institutes of Health in Bethesda found that melatonin secretion by the human pineal gland is suppressed by light, as it is in other mammals, but this suppression requires surprisingly much light in city-dwelling humans,[1] more than is ever encountered indoors: indoor light is like night to the human pineal. However, even indirect outdoor light is sufficient to

BRAIN CLOCKS

Spontaneous circadian oscillations are found in almost every kind of living organism. Possible exceptions include only those whose recent evolution transpired in the ocean depths or in deep caves, and those (procaryotes: bacteria and blue-green algae) whose cells have no distinct nuclei or mitochondria. Circadian oscillation is usual in higher single-celled organisms and in the isolated tissues of more complex organisms. Nonetheless, in both vertebrate and invertebrate animals it is usual for a part of the nervous system to be a dominant circadian pacemaker for the whole organism. Michael Menaker and co-workers have shown that in some birds (not all), that part is the pineal gland, rhythmically secreting a brain hormone, melatonin, on a schedule cued by light arriving through the roof of the skull. A sparrow's time zone can even be changed by exchanging its pineal for one from the new time zone.[2]

In rodents, too, the pineal secretes melatonin rhythmically, but under control from two clusters of neurosecretory cells, the *suprachiasmatic nuclei,* arranged to the left and right of the brain's midplane just above the crossing of the optic nerves. These paired clocks receive light/dark information from the eyes. Daily doses of melatonin in turn synchronize the circadian oscillation.[3] The suprachiasmatic nuclei play a similar role in monkeys. Human patients with lesions in that area of the hypothalamus show rhythm disturbances that suggest a similar role for the suprachiasmatic nuclei in humans. Their rhythm can be reset by light through the eyes, by electrical stimuli, by neurotransmitter analogues injected into the brain to simulate normal nerve discharges, and by melatonin. At least in rodents, removing the pineal gland enables the suprachiasmatic nuclei to adapt more promptly to a time zone shift. Perhaps an answer to jet lag will be found in some drug that suppresses the pineal while we adapt to a foreign schedule.

The pineal and its secretion, melatonin, are stimulated by psychomimetic drugs such as LSD, mescaline, and cocaine, and are suppressed by drugs

(continued on next page)

Overall neural activity →

21 09 21 09 21 09 21 09 21 09 21

Time of day

SCN
Subjective day

SCN
Subjective night

The upper graph plots overall neural activity versus time in the surgically isolated suprachiasmatic nucleus (SCN) of a rat in constant darkness. Activity rhythmically varies more than tenfold with a circadian period of 24 hours and 19 minutes. Below, autoradiographed sections through the rat's brain show metabolic activity by the radioactivity of bound deoxyglucose. A brain fixed during the active phase of neural activity (left) shows metabolic activity in the SCN; another fixed during the inactive phase (right) shows no activity in the same place.

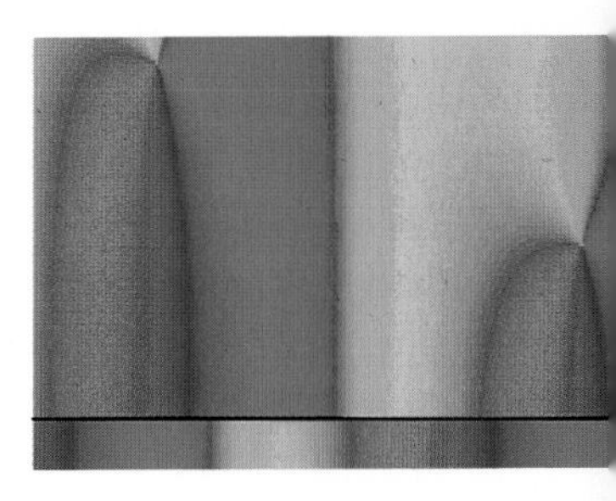

The ancestors of this cave salamander found little use for vision, so offspring with more and more rudimentary eyes were at no disadvantage. Might innate circadian rhythmicity have been lost too, by random mutation for many generations in temporal isolation?

now used in the treatment of psychosis. A widely used benzodiazepine antidepressant has recently been found to reset the circadian clock in rodents, possibly through its known action on a neurotransmitter that has been identified in the suprachiasmatic nuclei. All this suggests some connection between psychiatric diseases and irregularities of circadian coordination, especially between depression and irregularities of sleep.[4]

It is curious that humans need so much more light to suppress melatonin secretion than do other mammals. It would be of interest to know whether domestic dogs share this odd insensitivity. Could it be a consequence of exposure to indoor nighttime lighting through thousands of generations? If our circadian rhythms were affected by such dim illumination (as those of laboratory rodents are), then their derangement might feel like jet lag to individuals already stressed by other problems. Less sensitive individuals might suffer marginally less, and so might have fared marginally better in stressful situations affecting reproductive success. Have humans and their domestic pets been unwittingly bred for insensitivity to dim light? If so, then much greater sensitivity may still be retained in the Stone Age Tasaday of Mindanao or in the Maori or the Tierra del Fuegans, whose ancestors may have been less exposed to the putative circadian disruptions of nighttime light. Will that genetic inheritance be finally lost during this century?

---

immediately suppress the secretion of melatonin. This brain hormone has much to do with sleep and with the circadian clock; for example, daily injections of melatonin can entrain and synchronize the clock in rodents. If melatonin turns out to be a mediator of phase resetting in humans, then Lewy's observation will be of interest to anthropologists, to office and airport designers, to shift workers, and to travelers suffering from jet lag. Until a phase-shifting drug is selected and approved by the Food and Drug Administration, immediate heavy exposure to sunlight may become a pleasant medical obligation for travelers on sensitive business. Indeed, it may be just as important for nontravelers, who depend on daily synchronizing daylight to keep their internal rhythms properly timed. Failure to do so can mean daytime sleepiness and restlessness at night—common enough complaints. It may be significant in this connection that Daniel Kripke and collaborators monitored the timing and amount of light to which normal young men expose themselves in the course of an average day. Even in sunny southern California, the amounts are so surprisingly small and so diversely timed that

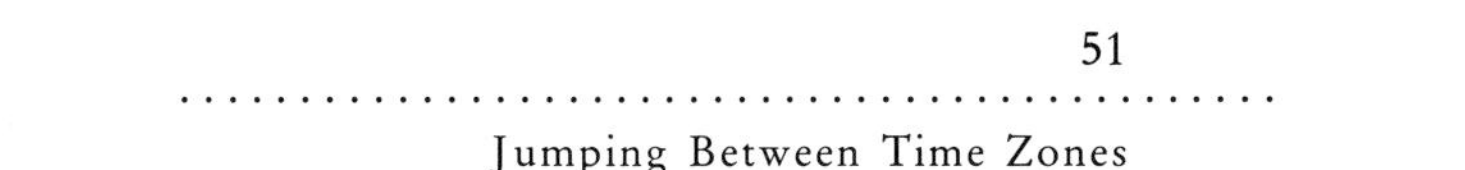

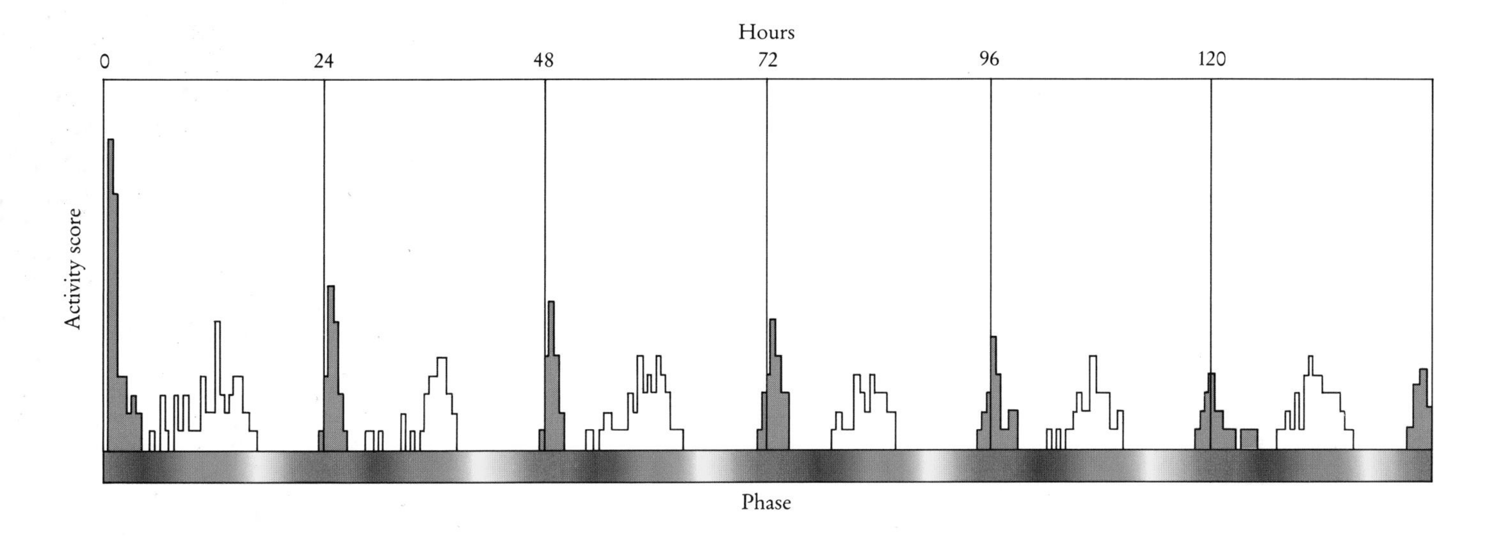

In a natural 24-hour light/dark cycle, the mosquito *Culex* is active around dawn and around dusk. During temporal isolation in a laboratory, this pattern persists with a period of about 23 hours. This plot shows the collective activity in a cage of 24 virgin female mosquitoes. The "dusk" activity is colored; "dawn" activity is not.

one may wonder how (or even whether) modern man keeps his circadian rhythms properly orchestrated.

An Experiment to Measure Phase Resetting

If circadian rhythms *are* kept properly synchronized by daily exposure to daylight, how is it done, and what does the answer reveal about the nature of internal clocks?

A simple experiment could be made with a light-sensitive organism whose circadian rhythm is conspicuous and steady in the absence of any cueing. The mosquito is a convenient subject. Like most other animals, mosquitoes have a circadian rest/activity cycle, in this case of about 23 hours' duration, persisting regularly even when the animals are caged in a room at constant temperature in constant darkness. Under these conditions of temporal isolation, a mosquito's activity can be conveniently measured by monitoring its sound: appropriate instruments record the number of one-minute intervals in each hour during which an annoying buzz is detected. In a cage of mosquitoes, this buzzing increases hour by hour, rising to a crescendo of high-pitched whine around sunrise and then again around sunset—or more accurately, at the hours when the mosquitoes *feel* that dawn or dusk should be near. In virgin females of the species *Culex pipiens quinquefasciatus* used by Eric Peterson,[5] the interval of "dusk" flights is more crisply defined, so he uses it to mark "phase zero" in the animals' circadian cycle.

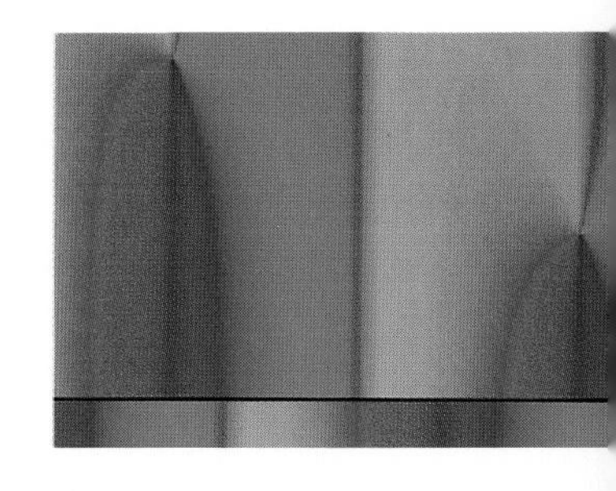

What would be a likely timing cue for the mosquitoes? To keep the experiment simple, the cue should be applied for a specific time interval—only a moment if that is sufficient, or many hours or days if that is needed. Peterson used white light of intensity comparable to outdoor daylight. Normally the mosquitoes are kept in perpetual darkness, but that darkness is punctuated once, and only once, by exposure to bright white light. This light pulse is administered in separate cages at different times in the circadian cycle. In each cage, the aftermath is monitored: the activity rhythm resumes, but with offset timing. The lower part of the figure on the facing page shows the prestimulus rhythm together with its extrapolation (shaded) as though no stimulus had intervened. (This could actually be monitored by leaving one cage unstimulated.) The upper part of the figure shows a typical poststimulus rhythm. As in the lower part, the shaded half is an extrapolation: only the unshaded parts are supposed to represent the time course of the actual experiment: a colored rhythm up to the moment when the environment becomes nonstandard (when the stimulus begins, at a phase in the cycle called "old phase"), and a colored rhythm resuming from the moment when the stimulus ends, leaving the cycle at a phase called "new phase". In a real experiment (the figure is an idealization), the rhythm may be somewhat deranged, even wholly suppressed, for a while after the stimulus, but it eventually returns to normal.

As the figure shows, the offset can be considered either an advance or a delay in the cycle. From observations of the end result, one cannot distinguish between those interpretations of the underlying clock process. Indeed, it may not be possible at all to describe that process in terms of advances and delays. To avoid making assumptions about the unobservable, it is convenient to refer to the offset as a change from an *old phase*, the phase of the rhythm at which the stimulus begins, to a *new phase*, the phase of the shifted rhythm extrapolated back to the moment the stimulus ends. (If the stimulus tapers to an indistinct end, it is convenient to assign a nominal end point one cycle period after the beginning.)

On the evidence that circadian clocks do entrain, we must expect the amount of the offset to depend on the old phase: a stimulus that had the same effect at all times could not be a useful timing cue. How does new phase depend on old phase? In the unshaded panel of the figure on page 54, the old phase is plotted against the new phase that results from a stimulus of fixed strength. The diagonal line shows where all the data would fall if the stimulus did nothing at all, for then new phase equals

Facing page: The effect of a timed light stimulus on freely cycling mosquitoes (the "dawn" activity bursts are omitted for simplicity). The lower plot represents the prestimulus rhythm, up to the yellow bar that marks stimulus time. The shaded portion after the yellow bar is what would have happened in the absence of the stimulus. The upper plot represents the poststimulus rhythm, beyond the yellow bar that marks stimulus time. In the gray idealized peaks it is extrapolated to time *before* the stimulus, just as the lower plot was extrapolated to time *after* the stimulus, always with a 23-hour period. As in the preceding figure, the colored phase rulers are positioned so that red coincides with "dusk" activity bursts. Read old phase on the lower color ruler at the end of the undisturbed interval before the stimulus (the beginning of the stimulus). Read new phase on the upper ruler at the beginning of the undisturbed interval after the stimulus (the end of the stimulus). The offset of the poststimulus phase ruler can be considered to be either a delay or an advance.

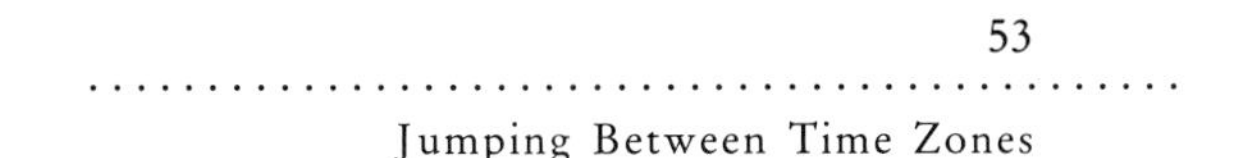

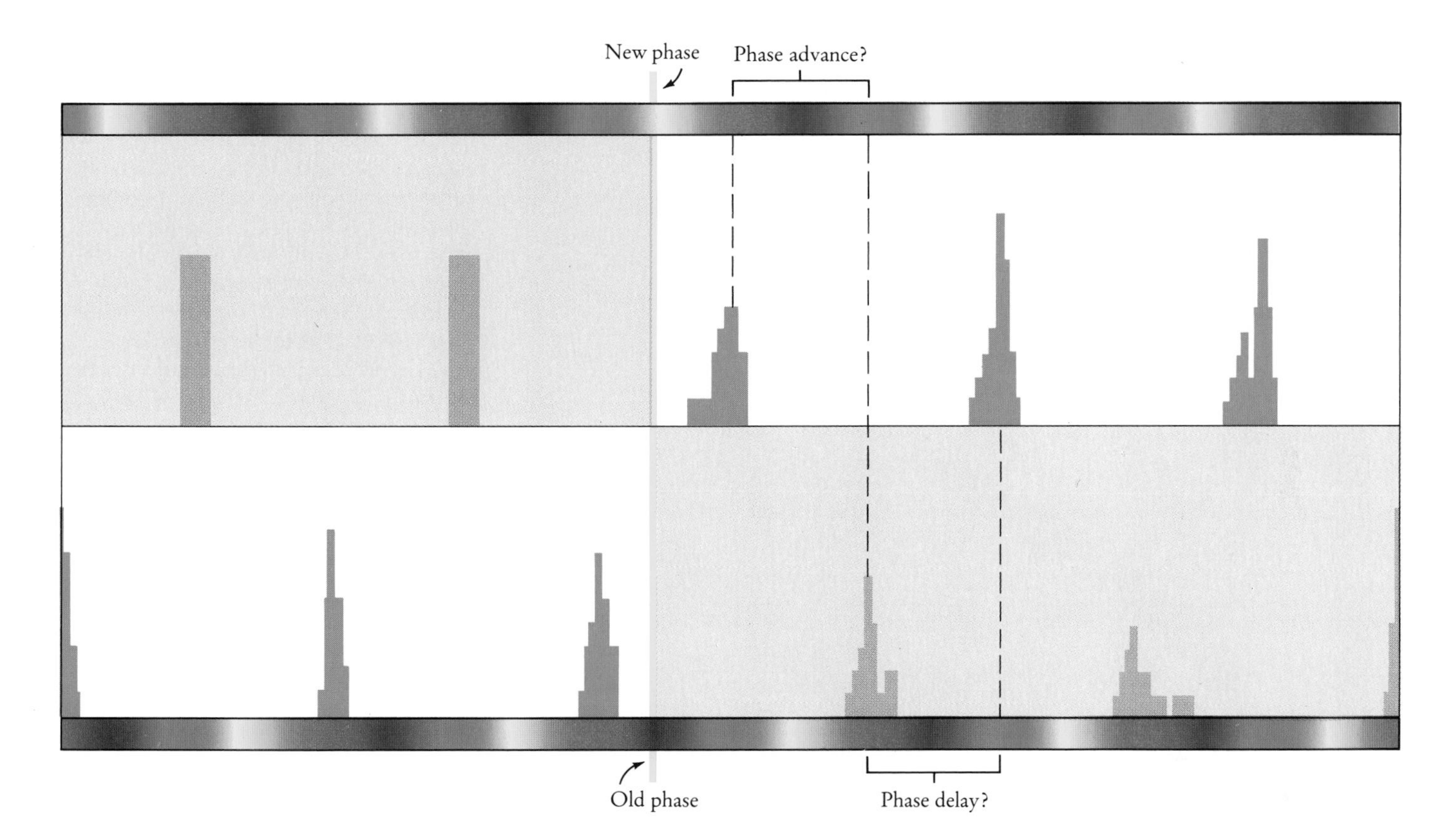

old phase (plus the duration of the stimulus, but that is constant). However, as you can see, the stimulus did have some cueing effect on this biological clock: when given at phase D, for example, the new phase became somewhat earlier; that is, timing was delayed (or conceivably, advanced). In contrast, after an exposure to light at phase A, the new phase became somewhat later; that is, the same stimulus applied at that different time resulted in a slight phase advance (or, again, perhaps it was a much larger delay).

The other three panels of the figure are deliberately redundant. For example, the old-phase axis is extended to the right to span two full cycles, as if conducting the same experiments through a second cycle of circadian oscillation. But the results imputed to the second cycle are simply replicas of the first. New phase is ambiguous, since a rhythm looks the same when shifted by any number of complete cycles. So every data dot might alternatively have been placed 23 hours higher, or 46, or 69 and so on; here this is indicated by replicating every dot one

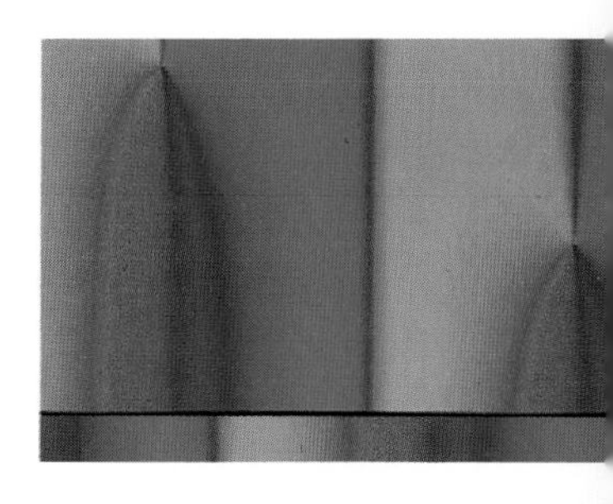

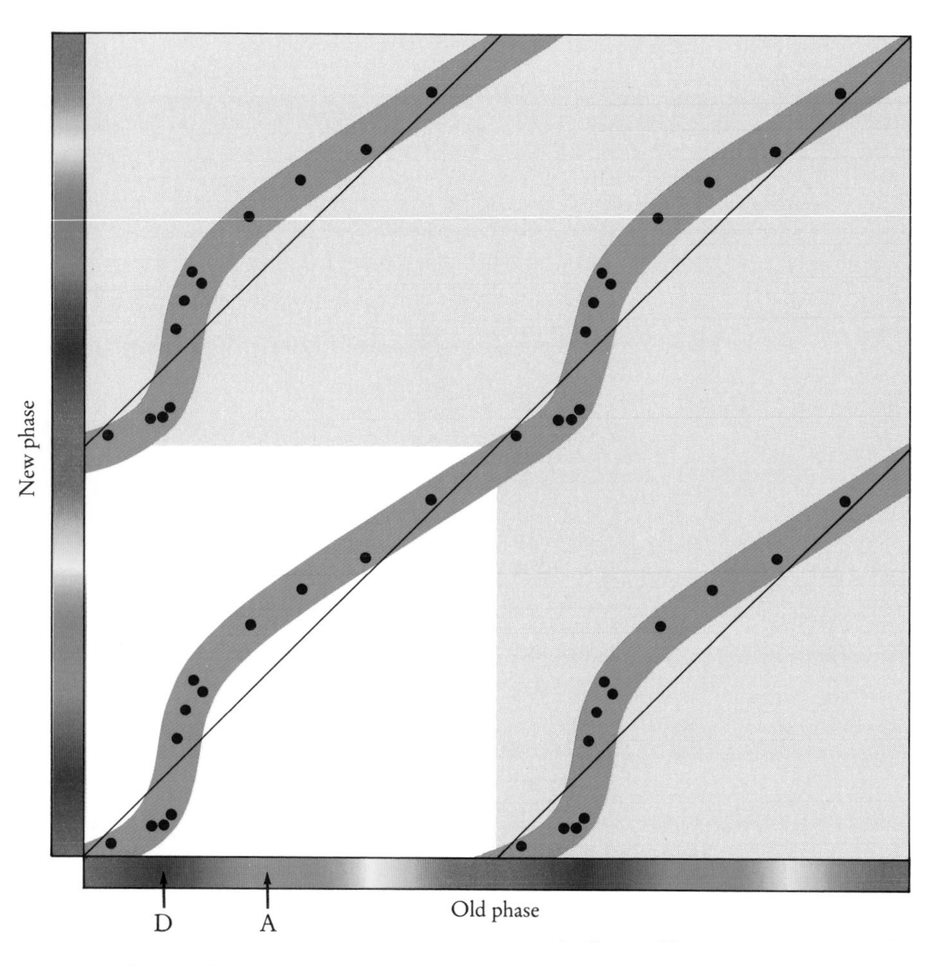

Mosquito activity rhythm: the new phase results of a 450-second light pulse applied at different old phases are repeated in the adjacent shaded squares to better display the continuity observed across the nominal beginning and end of each cycle. The diagonal line (new phase = old phase) depicts the result of control experiments in which the light pulse was too brief or too dim to change the phase. The results of 450-second light pulses depart from the diagonal, there being slight delays near old phase D and slight advances near old phase A.

cycle higher. The purpose of these redundancies is only to demonstrate how smoothly the trail of data dots runs across the boundaries of the unit square from the nominal end of one cycle into the nominal beginning of the next—as you would expect, since those boundaries represent only an arbitrary point called "phase zero" in a seamless circadian cycle.

Resetting Curves

How does phase resetting come about? Its results can be described without describing the process—a piece of good fortune, since no one yet knows the mechanism of a single circadian clock! The essential principle is merely that any cue (14 hours of daylight for example) acts differently on the internal clock's constantly changing mechanism, depending on when in the clock's cycle the cue starts. To a first approximation, the net effect of a cue can often be described as a mere offset of the internal clock's rhythm: one period after a cue began at some old phase in the internal clock's cycle (when, had the cue done nothing, the clock would be back to that same phase again) the clock is in fact found at some different phase, called the new phase. For present purposes there is nothing to distinguish this clock from an undisturbed clock that was already at that phase when the cueing stimulus ended one period ago. The net effect on the disturbed clock can thus be conveniently described as though the cue had instantly shifted the clock's phase from old phase to new phase. Many secrets lurk in the way the new phase depends on the old phase. Rediscovered in a dozen different specialist literatures, curves that relate new phase to old have gone by many names. A current favorite is "PTC" for "phase transition curve," a bit of a pun borrowing mystique from the unrelated subject of phase transitions in mathematical physics. To avoid letter codes and to stick to the traditional clock metaphor, we will use the simpler term *phase resetting curve* or merely *resetting curve*. The trail of data dots in the figure on page 54 outlines a resetting curve.

The importance of measuring the human phase resetting curve, using a standard dose of sunlight during a temporal isolation experiment, can hardly be overestimated. It is essential for dealing rationally with jet lag and other disorders of daily timing that result from shift work or seem implicated in recurrent seasonal depression, insomnia, and daytime grogginess. Appropriate attention to timing of outdoor activities may someday provide a noninvasive substitute for reliance on caffeine, sleeping pills, and antidepressants. Experiments are in progress to measure human phase resetting curves, and their results may be available even before this book is published. Then it should be possible to improve the efficiency of human entrainment to the daily light/dark cycle.

If the cueing stimulus (daylight) recurs regularly, resetting the clock each time in the way described by the resetting curve, and if the resetting curve itself remains unaltered all the while, then the consequences can be figured out mathematically. The results are well understood theoretically and are well correlated with experimental reality. It will not be our objective here to unravel this complicated story, but it deserves at least a few sentences in passing. Repeated application of the same stimulus at regular intervals can result in the clock's phase bouncing around chaotically forever. Or it can result in stable entrainment, the

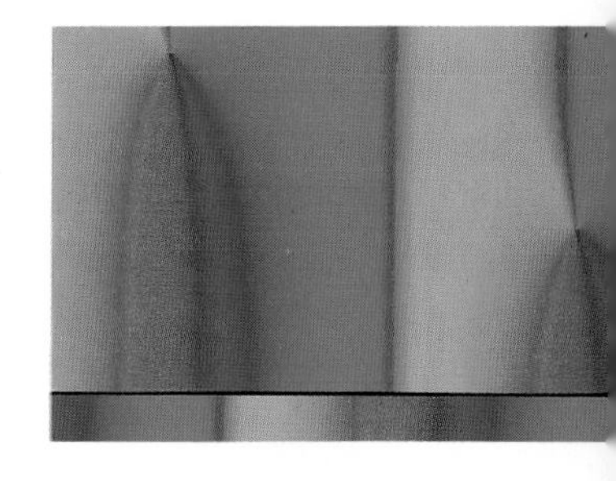

This mechanical clock indicates time from 0000 to 2400 hours around the rim of its face by numbers and by our standard phase ruler, the color wheel. The clock gets phase-advanced by the pull of a weight on its hour hand. At a different starting time it might be phase-delayed instead, or even be unaffected.

clock returning exactly to the same phase just as the resetting cue begins, every time; or alternately returning to one phase, then another, in a cycle of two cues; or in a cycle of four, or five, and so on. It all depends on the ratio of the clock's native period to the stimulus period and, of course, on the strength and nature of the stimulus and the nature of the clock. For clocks that immediately return to their standard cycle with only a phase shift, the results can be deduced from nothing more than a measurement of the resetting curve.

A Simple Clock Metaphor

What shapes might a phase resetting curve adopt? Let us begin by taking the clock metaphor quite literally. Imagine hanging weights from the hour hand of a mechanical clock, pulling downward on the hand, forcing all the attached geartrain so that the clock's entire mechanism changes speed. (The minute hand, geared to the hour hand, will move faster or slower and need not concern us.) Imagine that it is a TV news studio clock whose face reads a full 24 hours rather than just 12. At hours 0000 or 1200 the weights have little influence, but at 0600 they may cause some advancing, and at 1800 they may cause some delaying. When you take the weights off, the clock will not read the same time as it would have if let alone. What time will it read? That depends on the weight applied and its duration, and it depends on the hour when this interference was started. To simplify the discussion without changing the experiment, let us say that the duration of interference is exactly 24 hours, even if the weights are made too light to notice (are removed) during most of that time. Now make a graph.

Horizontally locate a dot on the graph by its reading when the weights were first applied (the old phase). Position the dot vertically by reading the clock at the end of this 24-hour interference with its free operation (the new phase).

If the weights are so slight as to have no effect, then new phase = old phase + 24 hours = old phase. This is the result at every choice of old phase, so we can quickly plot the entire curve as old phase is varied through a full cycle. The resulting resetting curve is a diagonal line.

Application of weights starting at 0600 will advance the new phase a little. If such weights are briefly applied at 1800, they will delay it a little.

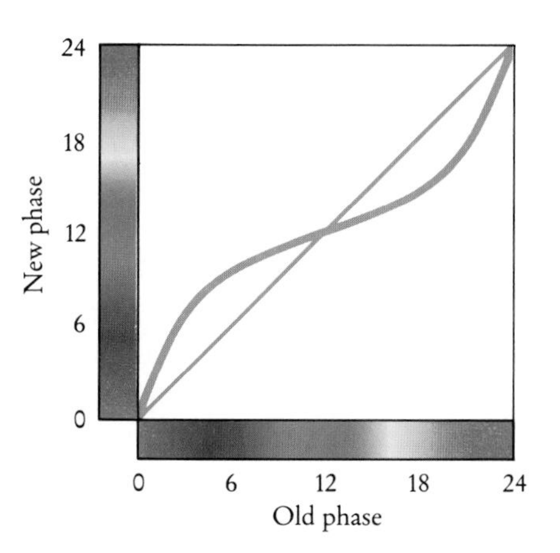

The mechanical clock's reading 24 hours after a weight was first applied (new phase) depends on its initial reading (old phase). If the weight was applied only momentarily, or was feather-light, then new phase = old phase (fine red), but if it exerted some substantial pull, then (thick red) new phase is somewhat advanced if the weight began pulling before noon, or somewhat delayed if pulling began after noon.

Two Kinds of Resetting

Circadian clocks are the same as mechanical clocks in three respects: their period returns to normal promptly after a stimulus, by a suitably chosen stimulus they can be forced away from the diagonal line showing new phase = old phase, and the amount of departure depends smoothly on the timing of the stimulus.

But they differ in another fundamental respect: that the resetting curve need not, as in the metaphor, be a mere distortion of the rising diagonal. This last point is a fairly recent discovery, 15 years old at most, and it has some surprising consequences. Most of this book is

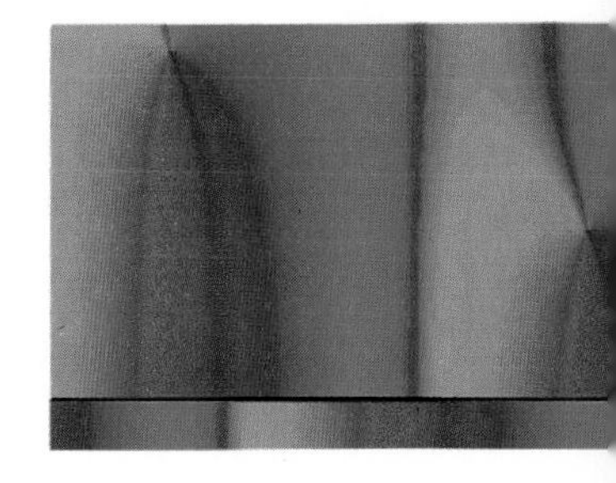

built around those consequences, so it is worthwhile to dwell on this point. Recall the mosquito's resetting curve, the result of cueing its circadian clock with a 450-second exposure to bright light. It resembles the resetting curve of the clock. But now compare the graph below, the result of a comparable exposure administered not to mosquitoes, but to fruit flies. The trail of data dots is not following the diagonal now: it follows a horizontal line. Is this fundamentally different? Whether a resetting curve follows a fundamentally horizontal path or a fundamentally diagonal path becomes clearer if the data are reduplicated in each direction, as though printing wallpaper. But a simpler way to represent the periodicity of old phase and of new phase is to just take *one* panel of

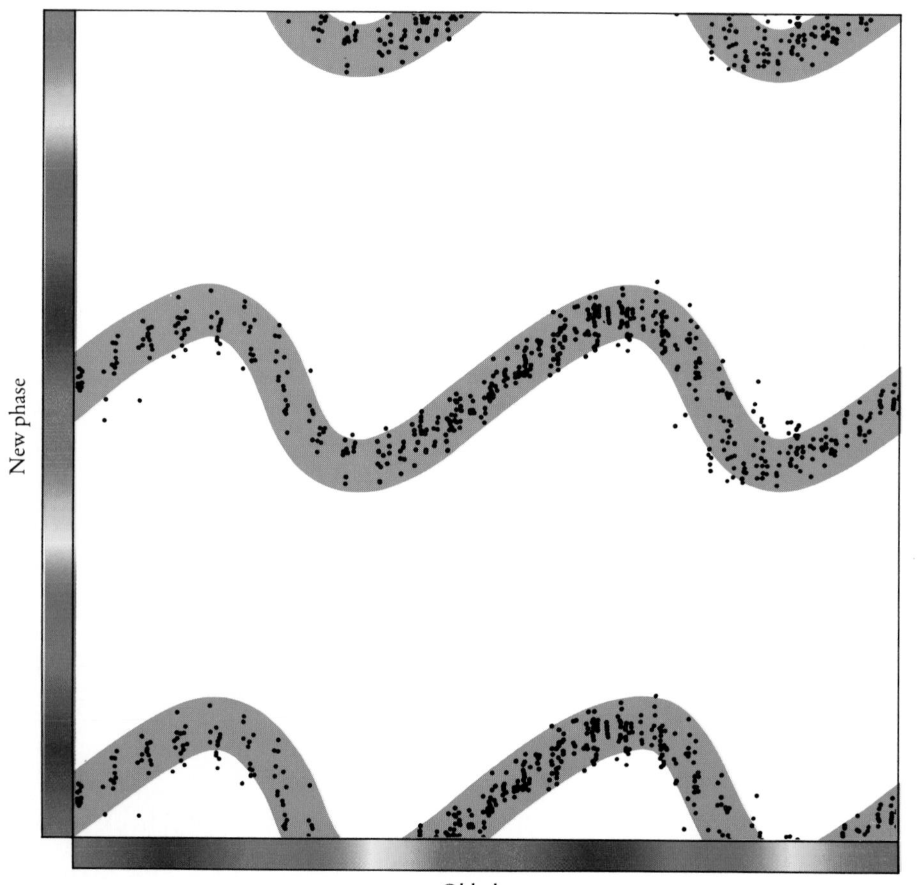

The resetting curve of a clock-timed activity in virgin female fruit flies, reset by exposure to blue light for two minutes. The experimental procedures and manner of graphing are similar to those described earlier for mosquitoes, but the data lie quite differently: the trail of data points is a horizontal wiggle instead of a diagonal wiggle.

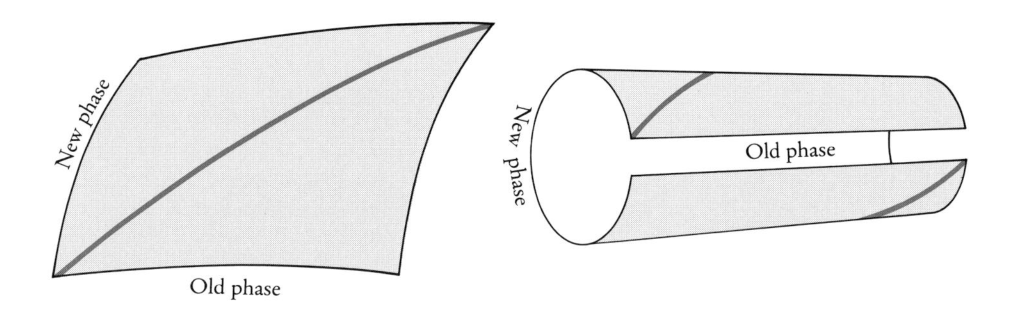

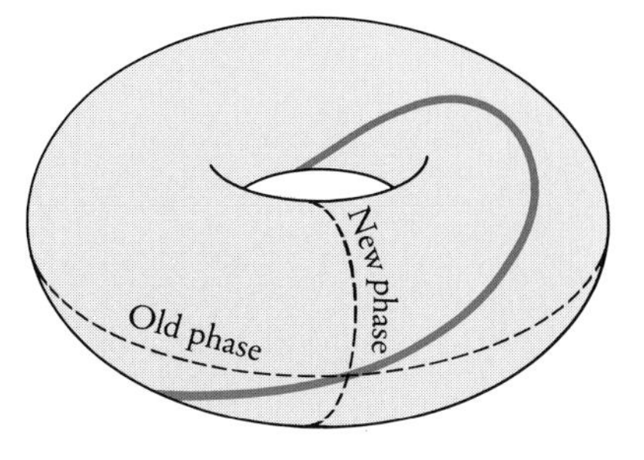

The left and right edges of a phase resetting graph represent the same new phase of the clock, and the top and bottom edges represent the same old phase. Here, the graph is folded into a torus. The new-phase = old-phase diagonal resetting curve (color) appears on the torus as a closed ring linking once through the hole.

the graph and paint it on a sheet of rubber so we can change its shape freely. Now roll it into a cylinder so the new-phase axis closes in a circle. The same must be done with the old-phase axis. The result is a sheet of graph paper that resembles the skin of a doughnut, a bagel, or a rubber inner tube—generically called a *torus*.

A torus may seem a funny kind of graph paper, but the choice is uniquely appropriate. First of all, we are going to be concerned only with topological features of the curve, those properties that are not affected by continuous distortions, for example, by stretching the graph paper. Second, both of the variables being plotted are phase values, points on a circle. The surface of a torus is two-dimensional and it can be ruled by circles in two perpendicular directions, one horizontal (running the long way around the hole) and one vertical (looping through the hole), to represent old phase and new phase.

What kinds of curves can be plotted on such graph paper? If a resetting curve drawn on it is to give a unique new phase to each old phase, then it has to be a closed ring of some sort: as we scan forward through one full cycle of old phase, the new phase changes continuously, returning to its original value at the end of the cycle (identical to the beginning). If new phase = old phase, then the curve (blue, in the above right picture) follows the diagonal on our coordinate grid, starting from the place where the two axes intersect, leading around the equator while looping through the hole, and returning to the intersection point to close the ring. Every kind of biological clock has this kind of resetting behavior in response to negligibly faint stimuli.

Now if a slightly more vigorous stimulus only slightly changes the timing of later events, the resetting curve is only slightly different, as shown by the trail of data dots from the mosquito experiment mentioned earlier. In our mechanical clock metaphor, even if the effect is not slight, the curve remains qualitatively the same: it is still a closed

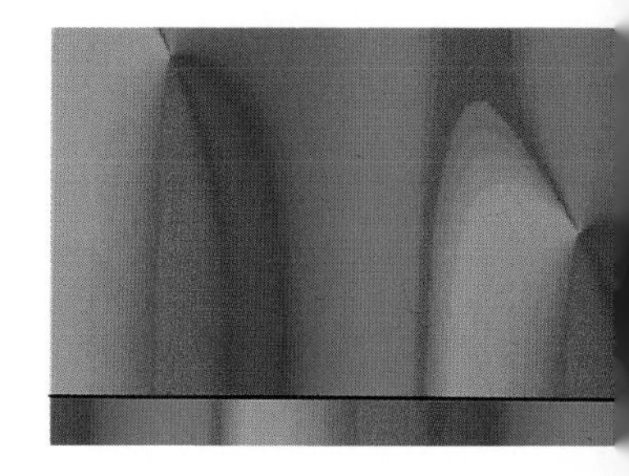

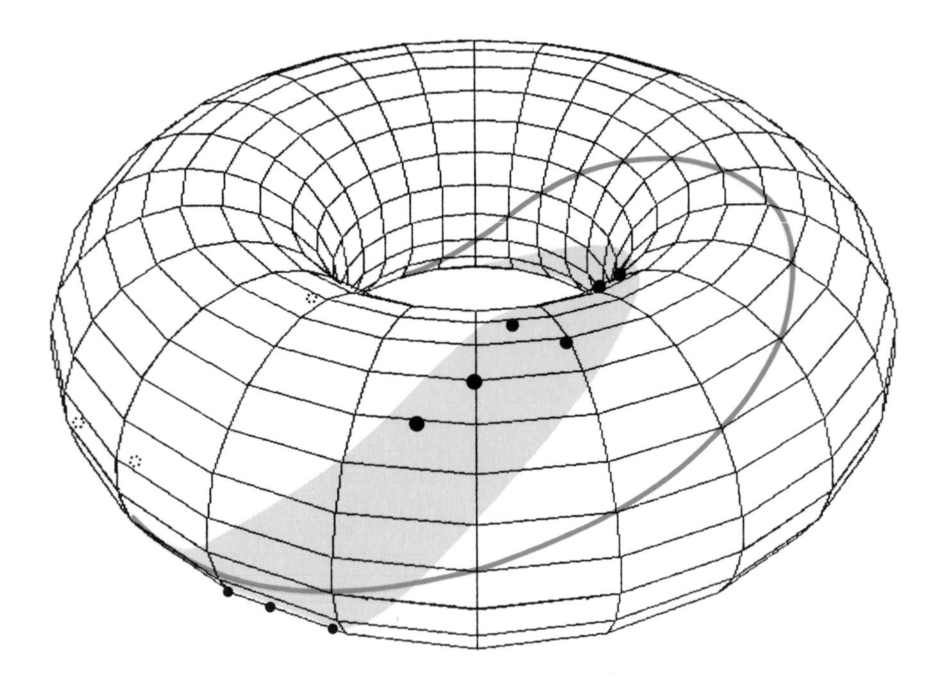

If made of rubber, the mosquito resetting graph could be rolled onto this torus. The diagonally wiggly trail of data makes a closed loop (color) linking the torus; a few data points are hidden where the loop is behind the torus. The thin colored line, also linking the torus, is the new-phase = old-phase diagonal.

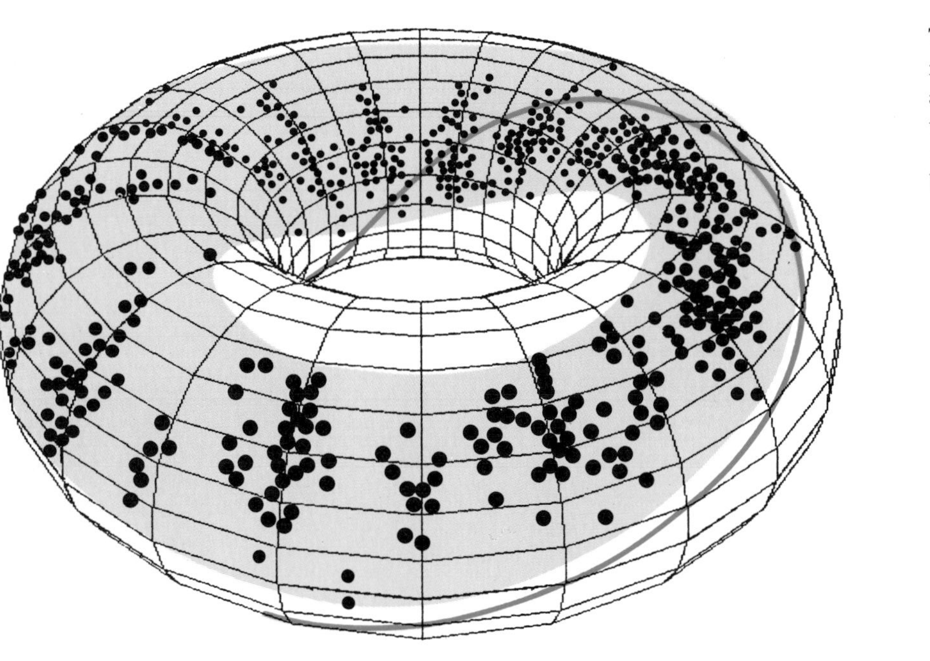

The fruit-fly resetting graph could be rolled onto this torus. All data points are exposed here: they lie along an unlinked ring (color). The new-phase = old-phase diagonal is shown in dark blue.

ring linking once through the hole. In terms of the graph on page 57, the heavy red curve is the fine red diagonal bent toward the horizontal line "new phase = 12"; heftier weights would bring it closer to that horizontal, but without severing its attachments at (0, 0), (12, 12), and (24, 24). In fact it is not hard to prove that this is the only kind of curve you will ever see as long as you deal exclusively with clocks (which I call simple clocks) that only speed up and slow down, at rates that depend only on environmental conditions (the current size of the stimulus) and the momentary phase of the clock. This type of resetting, represented by a curve that links the torus once, is called *type 1,* or *odd* resetting.

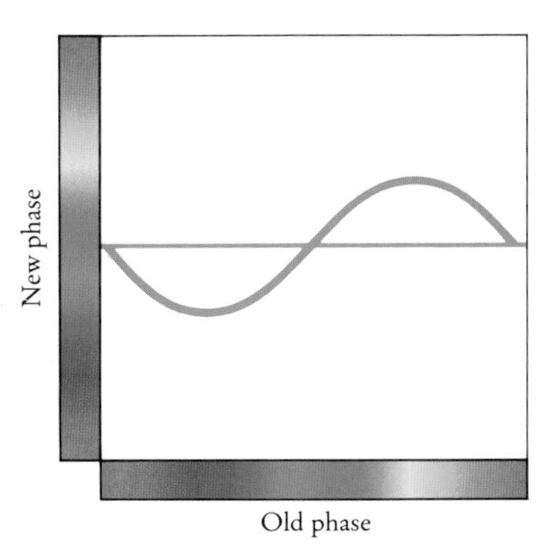

The fine red curve represents the limiting case of even-type resetting, to the same new phase regardless when the stimulus begins. The heavier curve is more representative of real measurements, such as the fruit-fly data on pages 58 and 60, but its topological linking number (0) on the torus is the same.

That is perhaps not too interesting. It seems to be just what was expected by every clock biologist who expected continuous results at all. What other kinds of resetting can there be? The resetting curve must be a closed ring, but must it link once forward through the hole? Apparently not: the lower graph on page 60 shows a resetting curve that closes in a smooth ring that is not linked through the torus. The exact shape and position of the ring encode the exact manner of cueing. We will not be too much concerned with these details, which differ markedly from one species to the next and from one kind of stimulus to the next: different durations, intensities, or colors of light, temperature pulses, administration of various drugs, and so on all inflict phase resetting in characteristic ways; the resetting curves differ in detail. The topological character of the ring, though—whether it links the hole or it doesn't—encodes something much more fundamental, which will be our main concern through the rest of this book.

In a limiting case of this new style of resetting, the stimulus simply resets to the same condition whenever a strong enough stimulus is applied: new phase is the same regardless of old phase. This type of resetting, represented by a ring that does not link the torus at all, is called *type 0*, or *even* resetting. Even or type 0 resetting is so called because the resetting curve on the torus links 0 times (an even number). It is sometimes also called strong resetting because it necessarily entails advances or delays (or both) exceeding a half cycle. Odd resetting is correspondingly called type 1 to remind that its curve on the torus links +1 time. It is also sometimes called weak resetting because it is compatible with very weak phase shifts and is always obtained in the limit of very weak stimuli. But the linking number is the important thing, not the magnitude of phase shifts. The size of phase shifts can be misleading: violent resetting can be part of a type-1 pattern if the new phase increases steeply through some range of old phase, but then decreases just as

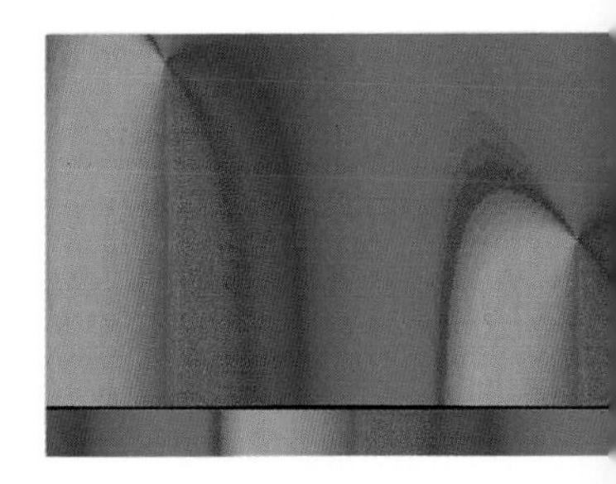

A PARABLE OF THREE CLOCKS

Here is a proof that the resetting of simple clocks can exhibit only the "odd" mode.

Take a sealed business envelope, cut off its ends, and open it into a cylinder. Mark each circular orifice with arrows pointing around the circle in the same direction to represent the normal sequence of clock phases through one cycle. The length of the envelope represents time during a stimulus of duration equal to one normal clock cycle. During this stimulus, the usual spontaneous advance of phase may be sometimes restrained or reversed, and sometimes accelerated. The paths of changing phase will be indicated by curves drawn on the envelope from some old phase at the left end to some new phase at the right end. For an impotent stimulus, the clock would continue to cycle normally, and the paths would be parallel helices winding once around the surface of the cylinder. For a stimulus that resets, finally, to the same new phase regardless of the initial old phase, all paths will converge to a single point at the right end.

Now suppose a stimulus results in a resetting curve that doesn't link the torus forward. (It fails to link or links any number of times backward.) If the curve is continuous and not dead flat, then there must be a region in which new phase decreases as old phase increases. On the left edge of the

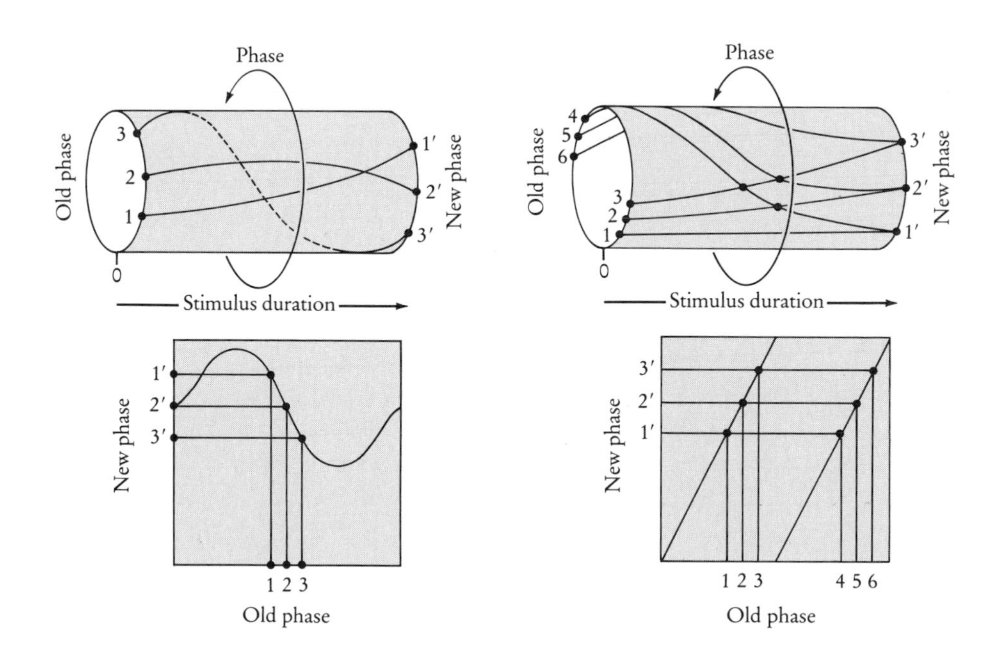

Left: A stimulus that causes new phase to wrap backward through zero or more full cycles must have a segment of negative slope (bottom panel). During the early stages of such a stimulus, new phase still depends on old phase in an increasing way; in fact, new phase = old phase in the case of a negligible stimulus. But during a prolonged or potent stimulus some range (1′ -2′ -3′) of new phase must come to decrease with increasing old phase. The unavoidable crossovers reveal that such a biological clock can have different internal states (and subsequent behavior) even though indicating the same time. A "simple clock" cannot. *Right:* An analogous thought-experiment reveals unavoidable crossovers also in the case of resetting curves that wrap two or more times forward through the full cycle of new phase.

cylinder mark three adjacent dots $1 < 2 < 3$ and trace their paths to the right edge so they arrive in the opposite order: $3' < 2' < 1'$. Notice that you can do it only if you let paths cross somewhere. At any point of crossing we have two clocks at the same instantaneous phase, exposed to the same stimulus; moreover both have been exposed to the same stimulus for the same duration. Yet their subsequent paths diverge. Clocks whose speed through the cycle depends only on the stimulus at the moment and on the phase at the moment must have the same speed when those two coordinates are the same; so they cannot diverge. Such clocks will therefore never produce data that link the torus backward or fail to link.

What then about a resetting curve that links the torus in the forward direction, but more than once? That implies that every new phase is reachable from two or more distinct old phases. On the cylinder, draw paths from three closely adjacent old phases $1 < 2 < 3$ at the left to three adjacent new phases $1' < 2' < 3'$ on the right, then a second set of paths converging to $1'2'3'$ from old phases 4 5 6 outside the range 1 2 3. Notice that the path from 5 must cross path 1 or path 3 to get to $2'$. Contradiction again. Such clocks will therefore never produce data that link the torus more than once in a forward direction.

That leaves only one possibility: a smooth closed ring linking once in the forward sense, the hallmark of odd resetting.

---

much through another range of old phase, so that the net change of new phase for each full cycle of old phase is still only one cycle.

In principle there could be other odd types (linking three times or once backward, for example) and other even types (twice linked in either direction, for example); it is not difficult to contrive mathematical models that exhibit these many resetting types. But none of the others have yet been encountered in biological experiments: type 0 and type 1 are the only two types of resetting yet found in the laboratory. Why only those two? There is no clear reason of adaptive utility to suggest that other types would have been ruthlessly weeded out in the course of natural selection. Is this a hint that circadian clocks evolved along two distinct lines? Or is it a hint about a fundamental mechanism common to all circadian clocks?

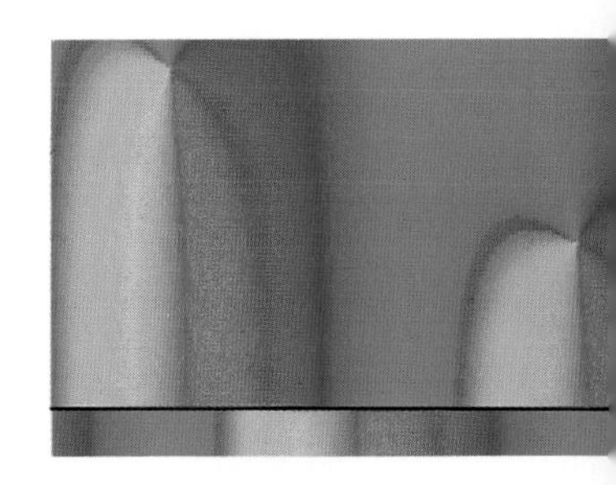

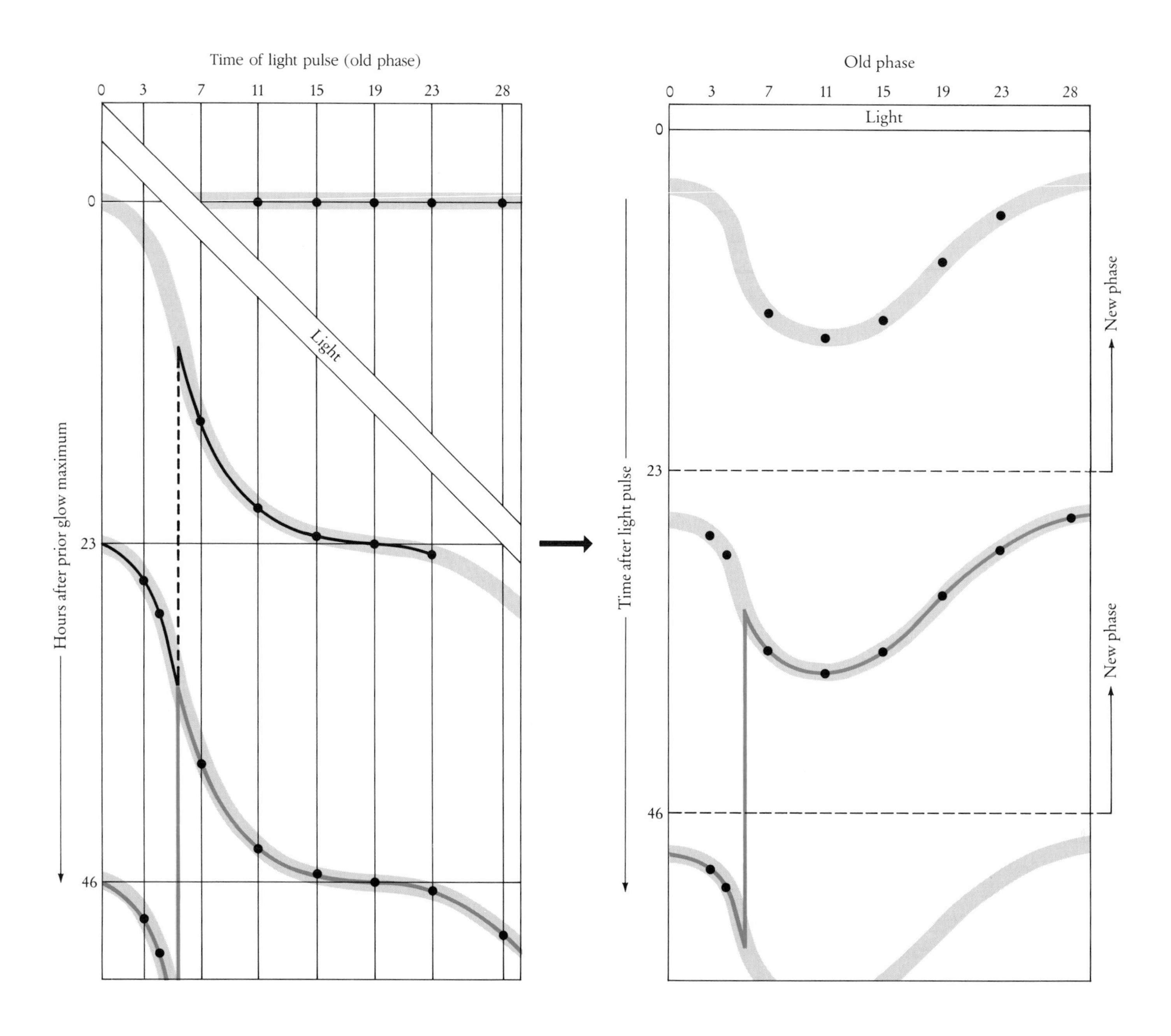
Time of light pulse (old phase)
0
3
7
11
15
19
23
28
Light
Hours after prior glow maximum
0
23
46
Old phase
0
3
7
11
15
19
23
28
Light
Time after light pulse
0
23
46
New phase
New phase

The First Even Resetting Curves

The truth knocks on your door and you say "Go away; I'm busy looking for the truth." And so it goes away. Puzzling.

ROBERT M. PIRSIG, *ZEN AND THE ART OF MOTORCYCLE MAINTENANCE*

Questions about the topology of resetting curves were first asked (and answered) less than two decades ago. As often happens in science, theory set the stage for recognizing phenomena that had already been recorded, but somehow could not be perceived until a conceptual home had been created for them in the minds of scientists. Like many conceptual novelties in science, even resetting escaped recognition for a decade after the data had been published. One of the first to tell us about it was a single-cell free-swimming plant bearing the awkward name *Gonyaulax* (pronounced *gone-ee-aw-lacks*). It is brown with chlorophyll and accessory pigments and makes good use of sunlight during the daytime. At night it goes about quite different metabolic business, such as reproduction by dividing in half. During the night it also emits a faint blue light, scarcely visible unless the cells are agitated mechanically. If jostled by a passing fish, each cell flashes brightly in irritation and we see a shimmering wake of bioluminescence. In the laboratory, even if there are no fish, a photomultiplier reveals that the cells still glow spontaneously, most brightly when their internal clocks tell them it is night (at intervals of 23 hours if they are isolated from the normal day/night cycle). In an experiment at Harvard University in 1957, J. Woodland Hastings and Beatrice Sweeney monitored the circadian rhythm of bioluminescence in little vials of seawater swarming with these remarkably precise clocks, collected from red tide patches in the Pacific surface waters near Santa Barbara and La Jolla.[6]

Hastings and Sweeney set up a lot of identical vials in temporal isolation and exposed each to a pulse of bright light, starting the irradiation of each vial at a different old phase in the cells' circadian cycle. In the graph at left, the moments of maximum glow before the light pulse are marked along a horizontal line at time 0 on the vertical scale: from left to right (increasing old phase) the light comes later after the maximum in each vial. During the light pulse, each vial's clocks are reset in a way that depends on the old phase. Subsequent glow maxima are advanced or delayed. The cells' response switches abruptly at about old phase 6, from a big delay in response to an even bigger advance, revealing a critical moment in the mechanism underlying circadian rhythmicity.

Left: Eight identical suspensions of *Gonyaulax* cells were exposed to three hours of light beginning at different old phases in their circadian cycle. The horizontal lines at 0, 23, and 46 hours show when the vials would have reached maximum glow if they had not been exposed. Data points show when the vials actually did reach maximum glow. One interpretation of the observed resetting would follow the black curve (including the dashed segment), deviating by advance or delay from the glow maxima of unexposed cells. Another would follow the (light colored) swath roughly parallel to the diagonal bar of light.
Right: The data are replotted to mark time downward not from the prestimulus glow peak, but from the stimulus itself (light). (This only requires shearing the graph upward by 45 degrees.) Now new phase at the end of the stimulus is indicated by the elevation of each dot above a horizontal line 23 hours (one cycle) or 46 hours (two cycles) and so on later. In this format, the original discontinuous curve through the data seems particularly suspect; this observation encouraged the search for data points on the putative vertical segment. None were found.

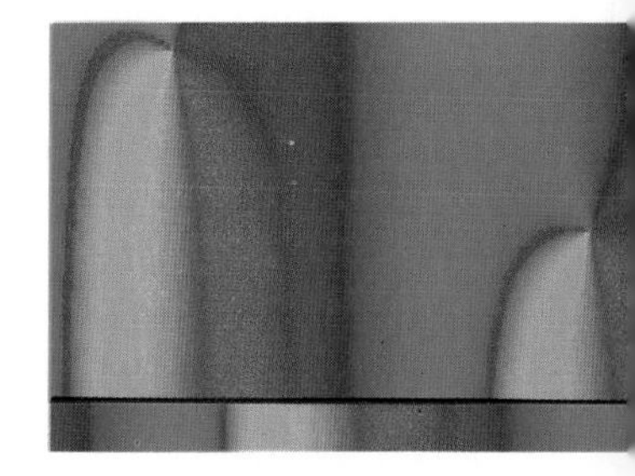

Or so it might seem. In fact, the advance and delay phase shifts called data might better be called interpretations. Consider an alternative interpretation. The glow maxima in the graph could also be linked by a smooth curve that wiggles along a diagonal parallel to the one marked "Light"; and another 23 hours later, and another 23 hours after that. There is no discontinuity. The same data are replotted in the graph at right, with only a trivial change in the layout: the "Light" bar is sheared up to a horizontal position so that, in each vertical column, time is measured downward from the end of the light, rather than from the preceding glow maximum. The smooth diagonal curves are thus sheared upward to become horizontal wiggles, each one periodic to the right at 23-hour intervals, and repeated vertically at 23-hour intervals. This amounts to a plot of new phase against old phase: new phase when the stimulus ends (or at 23 or 46 or 69 hours after that) is the number of hours already elapsed at that moment since the preceding glow peak. Again there is no discontinuity!

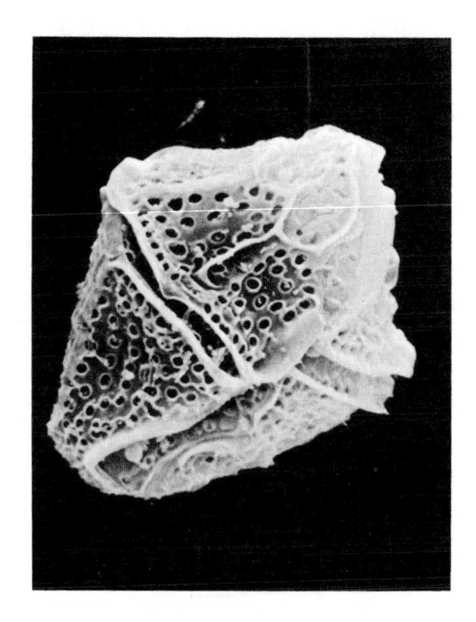

A single cell of the red tide organism, *Gonyaulax polyedra*, contains a circadian clock that regulates the timing of photosynthesis, cell division, and bioluminescence.

Where have we seen such a puzzle before? Perhaps at the international date line. There a traveler westbound from Greenwich with wristwatch delayed altogether 12 hours en route could meet a traveler eastbound from Greenwich with wristwatch 12 hours advanced. Their watches would agree exactly, though one traveler would say it is Thursday while the other said Wednesday. Sitting down to lunch together, they could eat with synchronously high appetites while arguing over the only difference between their perceptions of time: a matter of convention in naming whole days. In exactly the same way, the seeming discontinuity in circadian resetting data derives only from a convention of naming the glow cycles after the stimulus: of vials illuminated just before the "breakpoint," biologists said "the first glow peak was a half cycle retarded," while for the next vial they said "the second glow peak was a half cycle precocious," though both events appeared near hour 32 after the light.

A careful scientist would resolve this dilemma by pointing out that neither plot shows either continuity or discontinuity; actually, both show only a handful of data points, and any curve drawn through them is only the observer's personal interpretation. One can draw continuous or discontinuous curves as one pleases.

However, as years passed, more data points were collected and more laboratories made comparable measurements using other organisms, spacing their experiments very closely together. It became harder and harder to draw a plausible discontinuous curve. The graph on page 58, for example, shows data that were obtained in a similar way, from fruit flies, and plotted like the graph on page 64. Their continuity is now clear. Nevertheless, by the procrustean expedient of cutting this smooth curve into two parts and connecting them by a 24-hour jump devoid of data points, such observations were usually fitted into the traditional format of ostensibly odd resetting, thus preserving the appearance of advances and delays separated by a steep or discontinuous "breakpoint." One could then contemplate the genetics, biochemistry, and evolutionary meaning of that abrupt dislocation; but this exercise has never been rewarded by any clear conclusion or inference, and it still leaves us dismissing as coincidence the facts that the jump spans the rhythm's exact period and that the advance part of the curve at the top of the break has exactly the same slope as the delay part at the bottom of the break.

If this data-free break seems artificial, the alternative continuous curve leads to an even stranger inference. The next chapter develops that inference and shows how to test it experimentally.

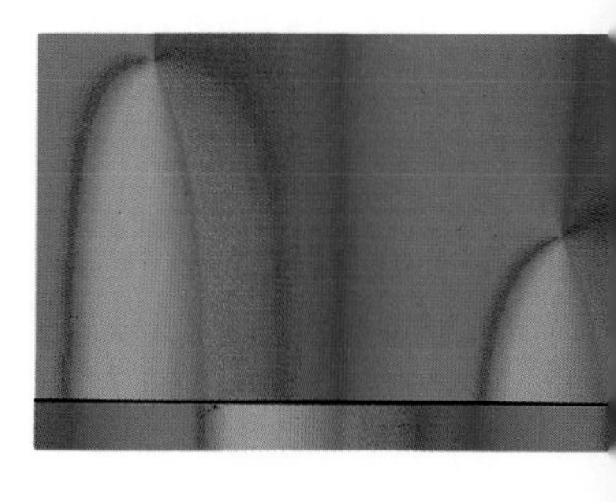

If timing runs through a full cycle smoothly around a ring (indicated above by the full cycle of colors around the edge), then within the ring there must be a region of discontinuous or ambiguous timing. The hueless point is the minimum expression of this anomaly, a phase singularity adjacent to all phases while itself at no phase.

Chapter Four

Convergence of Time Zones

The shop seemed to be full of all manner of curious things—but the oddest part of it was that, whenever she looked hard at any shelf, to make out exactly what it had on it, that particular shelf was always quite empty, though the others round it were still crowded as full as they could hold.

LEWIS CARROLL, *THROUGH THE LOOKING GLASS*

Smooth phase resetting is characterized by the curve of new phase values reached when a stimulus is applied at each possible old phase: that curve is always a closed ring on the toroidal graph paper appropriate for these measurements. In some organisms it links the hole once (odd resetting), and in others it doesn't link at all (even resetting). Perhaps this is a basis for classifying clock mechanisms or stimulus mechanisms.

But mustn't all organisms exhibit odd resetting, at least from very faint, very brief stimuli? In that case, new phase = old phase, the prototype case of odd resetting. Needless to say, this thought experiment works perfectly in reality with every organism tested. Conversely, in organisms originally characterized by the odd response, might their resetting be converted to even style by strengthening the stimulus? The plot on the next page shows a representative experimental result: by lengthening the mosquito's exposure from 7.5 minutes to 2 hours, the trail of data dots is not just bent farther from the diagonal (where it necessarily started, with 0-second exposures), but is in fact converted from odd to even topology. The sixteenfold change of stimulus strength evokes the change dramatically, but it can be evoked by a much smaller change, from the most prolonged exposure that causes odd resetting, to the adjacent briefest exposure that causes even resetting.

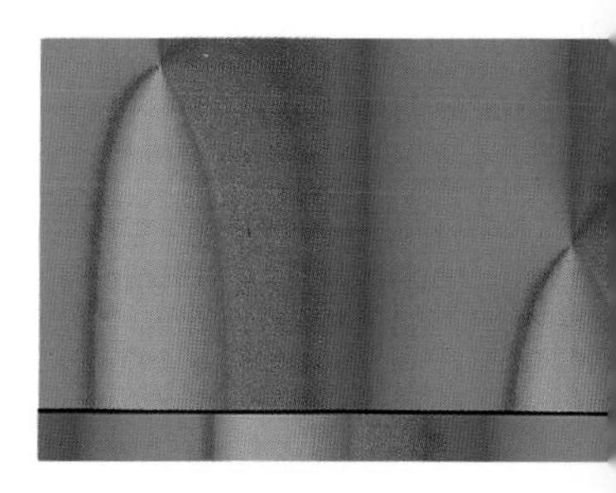

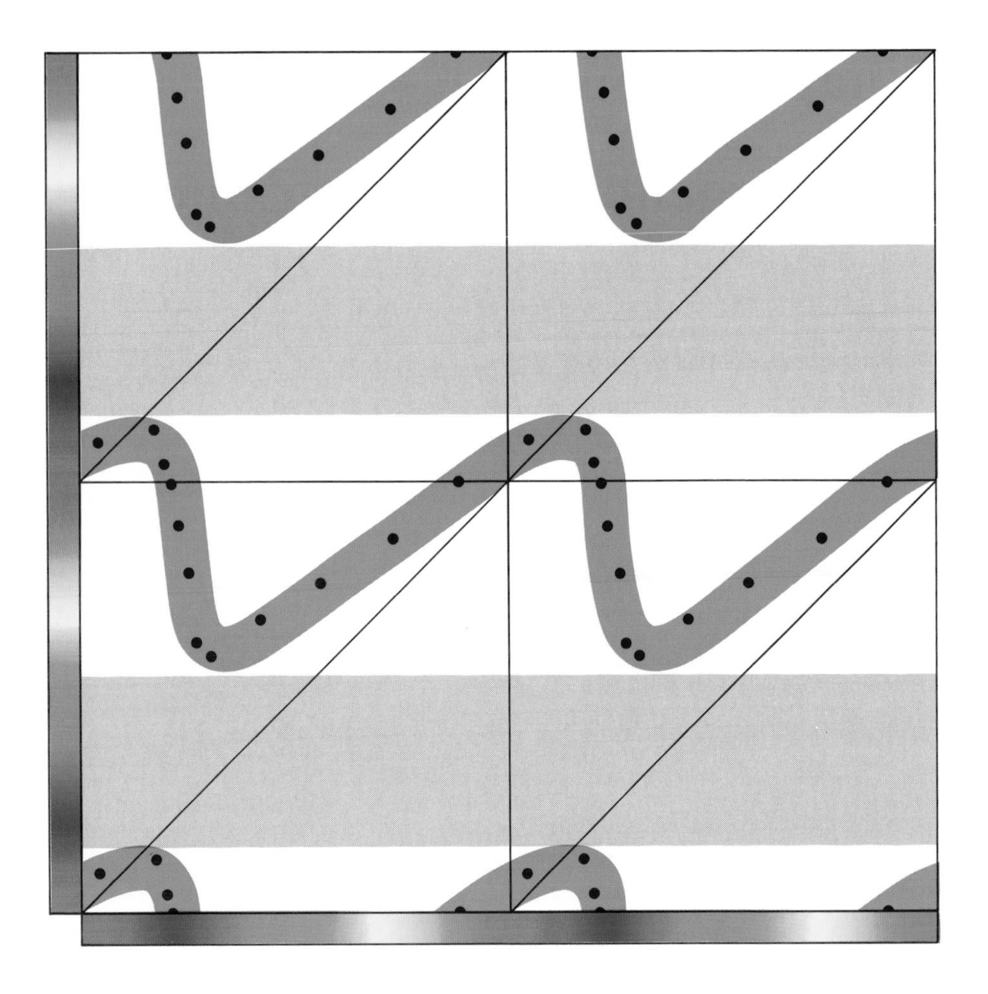

A resetting curve for mosquitoes. Everything is the same as in the plot on page 54, except that the light lasted for 2 hours instead of only 7.5 minutes. The trail of data points is now a horizontal wiggle instead of a diagonal wiggle. A 10-hour wide band of new-phase values (gray) is never crossed.

In one stroke the idea is thus demolished, that topologically distinct styles of resetting are due to distinct kinds of clock mechanisms or types of stimuli. Distinct styles of resetting turn out to bear witness not to diversity of mechanisms, as might have been assumed at first, but rather to universality of behavior! But now we are presented with a new riddle: If you reset almost any circadian clock, starting from almost any old phase, you get a new phase that depends smoothly on stimulus size. Yet the change from odd to even resetting, from linked to unlinked rings, cannot be made smoothly. How do you deform a ring linked through the hole into one that merely lies on the surface, unlinked? You don't, without a scissors. Neither does Nature. In *Culex* mosquitoes, for example, when the stimulus exceeds a certain critical size, the resetting curve changes abruptly from one that passes through all values of new phase to one that *avoids* a considerable range of values. For some "vulnerable" old phase, then, when the stimulus exceeds the critical size, the value of the new phase will jump.

The word *jump* connotes discontinuous change, but surely this is an exaggeration; except for molecular or quantum mechanical events, change is mediated continuously in time, even if its continuity is evident only on a very short time scale. True discontinuity—a finite effect from an infinitesimally small change—is not only rare, but also unobservable: it is impossible to be sure that the cause was truly smaller than any finite amount. Yet there seems no alternative to the inference that new phase must change by a finite amount while old phase increases only infinitesimally—a true discontinuity. How can we be so sure? Simply because the problem is evident even without careful measurement of quantities; it is a topological problem. A ring that links the torus must somehow be deformed into one that does not. There is simply no way to do this gradually. At the critical stimulus size the ring must be cut somewhere, pulled around the thickness of the torus, and reconnected. Topology unaided by empirical science cannot tell us what to expect near such a singular point, but it does tell us to go look: something physiologically singular, something other than mere rescheduling of the same old rhythm, must happen. That singular something had never been even hinted at in any experiment 20 years ago; yet from only well-behaved experiments, each resulting in a clear phase shift that depends smoothly on experimental conditions, the existence of something quite different and never before observed could be rigorously inferred. More surprising still, topology provided a recipe for finding a singular point in the laboratory. This *singularity trap protocol* made feasible the first experiments designed to detect the phase singularity in a biological clock.

How to Trap a Singularity

Up to now, the vast majority of mathematical models for biological clocks remain untested or untestable, so it is fortunate that, in some ways, they are also unnecessary: essential biological properties of living clocks can be derived by thinking logically about resetting data. Furthermore, this kind of reasoning can be put on an intuitive geometric footing by the use of color-coded phase diagrams. Such diagrams can

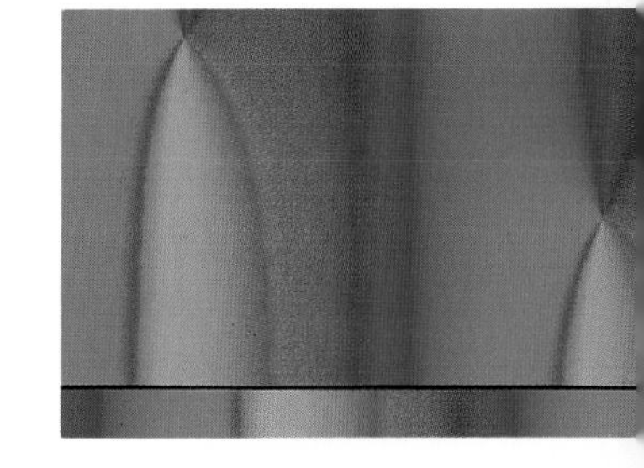

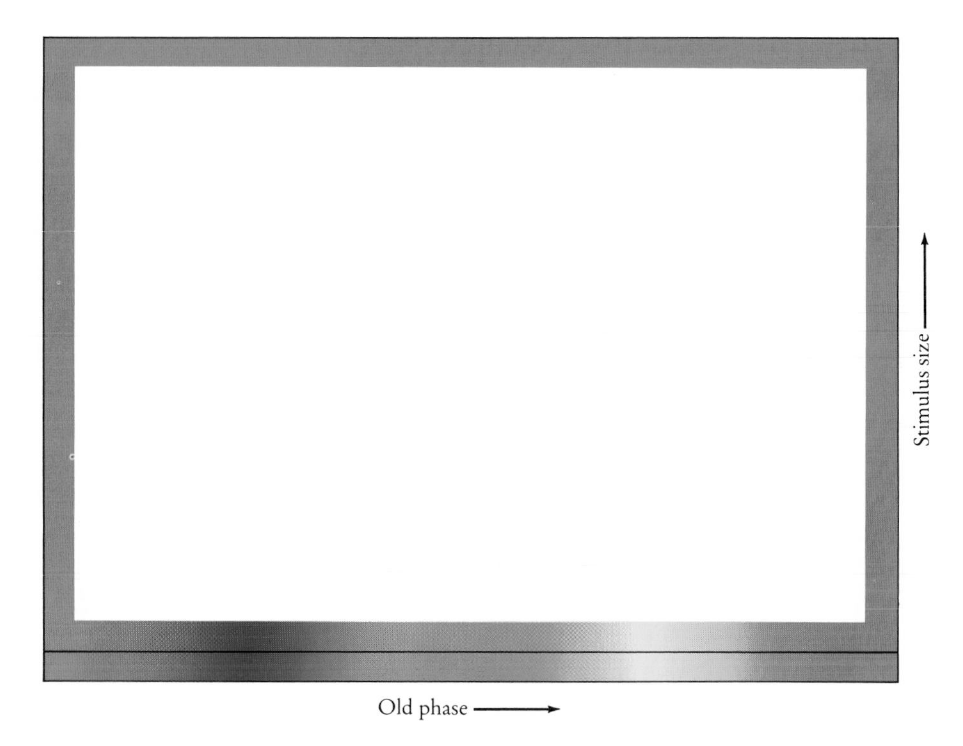

A stimulus is characterized by some size (a measure of duration or strength) vertically, and by the old phase when it begins, plotted horizontally. Here, the resulting new phase is color coded along the edges of the plot.

show directly the surprising implications of the discovery of even resetting, without laborious attention to topological proofs.

Begin by simplifying. Consider idealized, extreme cases of odd and even resetting: in the odd limit, the new phase is the same as the old phase; in the even limit, the new phase is the same regardless of old phase. It will be convenient to call that particular phase *zero.* Thus at old phase zero, after either a hefty stimulus or an insignificant one, the new phase will remain zero; let us suppose for now that it remains zero also for all stimuli of intermediate size.

What about stimuli of intermediate size at other old phases? It seems reasonable to assume that changing the stimulus size and timing ever so slightly will change the result only slightly. Without that postulate of continuity, nothing more could be said—since stimuli cannot be repeated perfectly, any measurements made in the past would have no bearing on expected results of future experiments. Now make a rectangular graph on which to plot each possible combination of stimulus size and old phase: locate each possible stimulus horizontally according to the old phase when it begins, and vertically according to its size.

At each point in this rectangle we want to plot the result—the new phase. We already know the results when the stimulus is very big (constant new phase), and when it is very small. We also know that, at phase zero, the stimulus has no effect regardless of size. In other words, we know the results all around the border of the rectangle, and we can plot the new phase values as colors: along the bottom, new phase = old phase, so the coloring increases from red (at phase zero) to violet to purple to blue to green to yellow to orange and back to red; the other three sides are uniformly at phase zero = red.

Now imagine extending the color plotting of new phases into the interior by trying stimuli of every size at every old phase. In terms of the color code, our assumption of continuity means that each color must blend into colors adjacent to it on the phase scale. For example, yellow may not blend into blue without an intervening area, however small, of either green or the sequence orange-red-purple-violet. Thus, the red upper border may grade into orange or purple as the fringe of color extends inward, while from the floor tongues of violet, green, yellow—every hue—reach upward to fill the interior. Given the assumption of continuity, they all have to merge somewhere. Try it with a box of colored pencils. Every pattern will somewhere contain a gray jumble of all hues.[1]

At some point, some combination of stimulus size and old phase, there is no well-defined new phase, but points very nearby are tinged with every hue. That point is a phase singularity.

Do you have a feeling of déjà vu? In Chapter 1, time zones converged to a timeless point at a geographic pole and tidal contours converged to amphidromic points. Resetting the rhythms of biological clocks moves them also to new time zones. As with time zones and tidal contours, there is a closed path (the border of the rectangle, in this case) along which every zone is encountered once, in order. So the interior of that path must contain at least one point of ambiguous zone.

There is a topological theorem about this inescapable phaselessness, called the *non-retraction theorem.* It concerns an attempt by all the points of a "manifold" (here, the colored rectangle of stimuli) to retract to the boundary (to adopt the coloring of a point on the fringe of the rectangle) without puncturing or tearing the manifold (without losing continuity with its neighbors). The theorem proves that it cannot be done: at least one point must remain indecisive.[2]

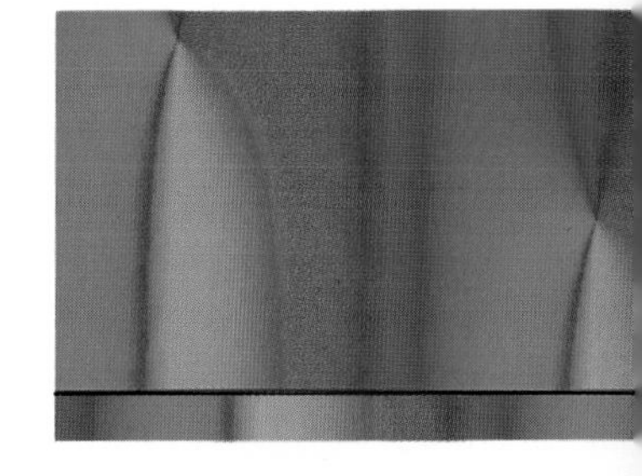

No matter how the rectangle is filled in under the assumption of continuity, the colors must converge somewhere to a hueless center.

This seems to mean that there must be a singular combination of stimulus size and timing such that the aftermath will not be any reproducible resetting of phase. What then may result? A different kind of rhythm? Arrhythmia? Quiescence? And at nearby points, unpredictable resetting of the prior rhythm? Only experiment can answer these questions. A few experiments and answers will be sketched in the next chapter.

Meanwhile, the colored rectangle may serve as a guide in the design of efficient experiments to discover the singularity. A particular point in a plane is like a needle in a haystack, so a moment of study could save months of redundant experiments. Imagine performing sets of resetting experiments alternately along horizontal rows and vertical columns, as shown in the picture on the facing page. Along horizontal strip K, we already Know the results (no stimulus, so new phase = old phase). With real organisms, it is too much to expect that the other three sides of the rectangle will all have the same color, but along vertical strip U, repeated on left and right one cycle apart, the results are Unimportant,

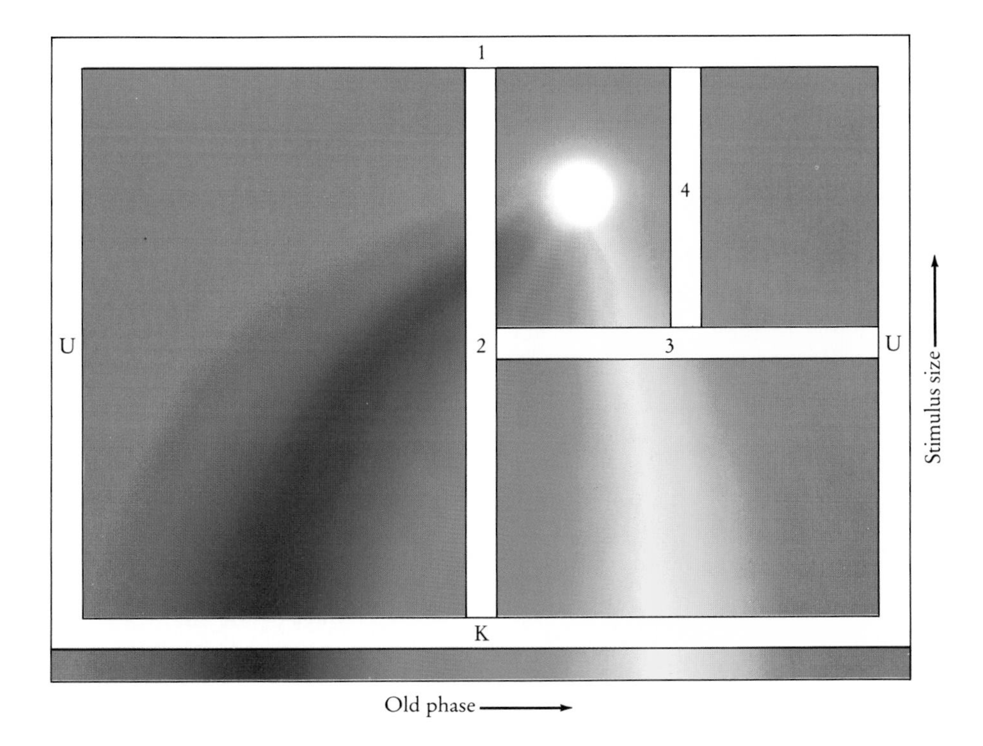

The hueless point can be located by enclosing it with rows and columns of experiments that progressively subdivide the plane without opening the color cycle. This method of successive bisections is a *singularity trap*.

just so they change continuously. We begin the experiments along horizontal segment 1, using a stimulus potent enough to elicit even resetting; thus, even if the color changes as old phase increases, it reverses itself before old phase has scanned a full cycle horizontally from U to U. Notice now that a tour around the whole rectangle takes us through a full cycle of hues.

Next, divide the rectangle in two by a vertical line of experiments (segment 2), each measuring the new phase resulting from a stimulus of increasing size, applied in separate experiments at the same old phase in the cycle. Check each resulting rectangle to see whether its edge retains a full cycle of hues. (Theorem: one or the other must.) If not (left), forget it. If so (right), subdivide it by a horizontal line of experiments (segment 3) at fixed stimulus size but covering a range of old phases that splits the chosen rectangle into upper and lower parts. Check each part as before, choosing the one with a full color cycle to further subdivide by vertical segment 4. Eventually your box will be as small as you please (or as small as reproducibility of experiments allows). Within these narrow

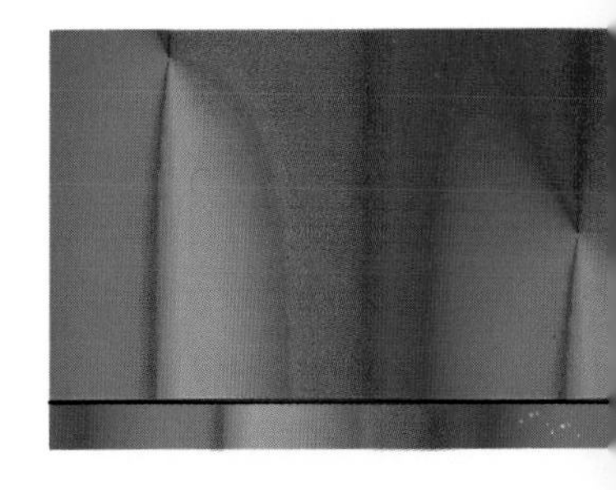

confines, the stimulus size and timing scarcely varies, yet the resulting new phase (hue) still ranges freely over the whole cycle. This box apparently still contains the singularity. Its location reveals the vulnerable phase and the critical stimulus size.

This singularity trap protocol made feasible the first experiments designed to detect the phase singularity in a biological clock. The experimental subjects were the fruit fly's circadian clock (Chapter 5) and the alternating-current energy metabolism of yeast cells (Chapter 7).

A Three-Dimensional Graph

We have reasoned our way quite abstractly to the conclusion that there are time zones in internal clocks. From idealized limiting cases of odd and even resetting, we saw on a colored diagram that these time zones must converge to some timeless center, much as they do on the globe. The argument depends on seeing each phase of the cycle as a hue on the color wheel. But since we intend these colorful mathematical fantasies only as design work toward real experiments, it behooves us to reformulate the expected outcome in terms that are easier to compare with what we shall see in the laboratory. Here we briefly become crystallographers—not of chemical substance, but of biological time.

What do the data consist of? Not phase values directly, but event times from which we must infer them. How is old phase actually measured? It is the fraction of a cycle elapsed from the beginning of the cycle—conventionally marked by some monitored discrete event, such as waking from sleep—until the stimulus begins. A cycle is 24 hours long, or whatever its native period is. To measure new phase, time is marked from the application of the stimulus; the new phase is the fraction of a period from the time of the next monitored event to the time one period after the stimulus. In practice, it is convenient to measure the new phase not one period after the stimulus (when the organism may still be disturbed), but several periods later, when everything has been routinely rhythmic for a while. For reasons of tidier theoretical interpretation, it is also customary to measure new phase at a whole number of periods after the stimulus *ends,* rather than after it begins, but the difference is inconsequential. It is obviously inconsequential if the stimulus is brief, but even if the stimulus is prolonged there is no difference between the two definitions if we artificially define the stimulus as "influences during one full cycle after they begin" (during most of which interval the influence has usually been turned off anyway).

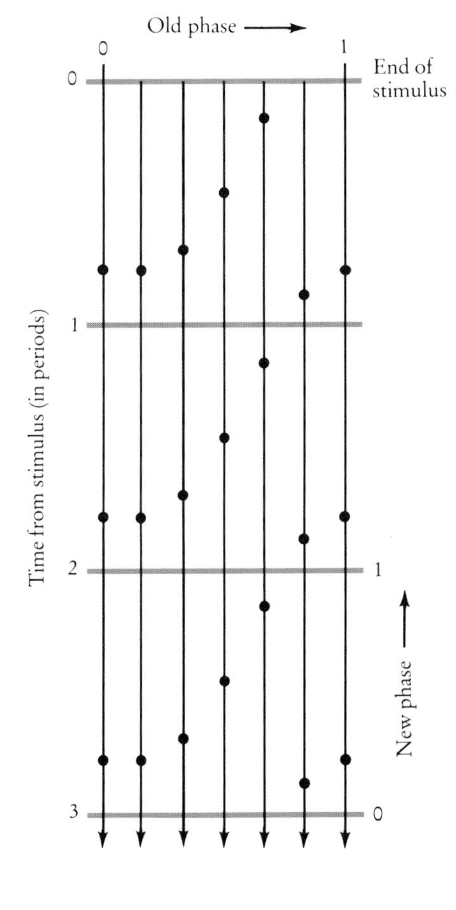

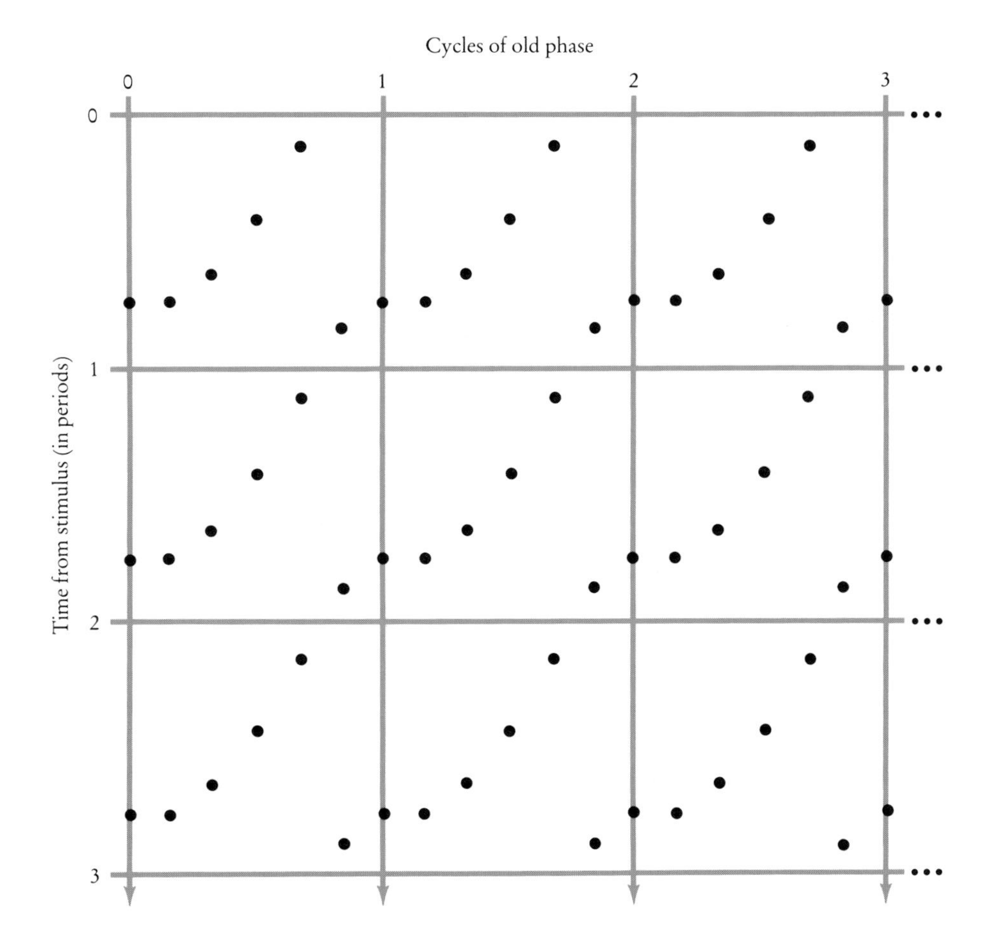

Left: Data from a series of resetting experiments is plotted on a set of vertical downward time axes, each beginning at the top when a stimulus ends. The old phase at which the stimulus began is used to position the time axis laterally. On each time axis, the monitored event marking phase zero recurs periodically from top downward. The horizontal lines mark multiples of one period after the stimulus. The distance of each dot from the next horizontal line measures the new phase. *Right*: If successive cycles of old phase are shown separately, the plot takes the form of a two-dimensional crystal lattice.

If cueing stimuli are applied in more than one cycle of the clock, each cycle of old phase may be presented separately along the old-phase axis. In that case, the plot of data resembles a two-dimensional lattice, repeating periodically in both dimensions. The horizontal period reflects the prestimulus rhythmicity of the circadian clock: whether we cue it now or exactly one cycle later when it is back to the same phase, the results should be the same. The vertical period reflects the poststimulus rhythmicity of the circadian clock: after resetting, it cycles in the usual way (except perhaps for an interval of "postoperative recovery" after the stimulus).

Connecting the data points in such a lattice produces the familiar resetting curve for a stimulus of certain size. Typical curves for stimuli of various sizes are shown at the right. The top example shows the limiting case of odd resetting, in which the stimulus has no effect, so new phase = old phase. The bottom example is the other extreme, in which the rhythm is reset to the same new phase regardless of when the stimulus is given. In proceeding from smaller to larger stimuli, the odd resetting curves gradually bend but still stay skewered on the diagonal. Beyond a certain stimulus magnitude, the curves appear in even resetting form: skewered on horizontal lines. How is the transition managed? There is an easy way to see, and the vision is a little surprising.

Imagine making a horizontal stack of lattice sheets arranged front to back in order of increasing stimulus size. New phase repeats vertically, old phase repeats to the left and right, and stimulus magnitude or duration increases to the rear. Each sheet in the stack is biperiodic like a tiled floor, so it is natural to call the entire biperiodic object a *time crystal.*

In the time crystal, the resetting curves fit together to make a wavy surface called the *resetting surface,* which shows how new phase depends simultaneously on the old phase and the size of the stimulus. Trace the resetting surface where it intersects the walls of a unit cell. Along the front wall, at stimulus size zero, new phase = old phase: such stimuli inflict no phase shifts. So we rise up through one cycle of new phase as we traverse one cycle of old phase from left to right. Going deep into the background now, turning left to hold the old phase fixed while increasing the stimulus, we find new phase changing in some way (the exact way won't matter for what we are about to find out). Now at the back wall of the unit cell, turn left again, going through one cycle of old phase while holding the stimulus fixed at maximum. This even resetting curve is skewered on a horizontal line: new phase merely wanders up and down, but returns to its original level as old phase traverses one full cycle and returns to its original value. So there is no net change of phase as we follow the surface along the back wall. Now finish the circuit by turning left to come forward again, along the left wall of the

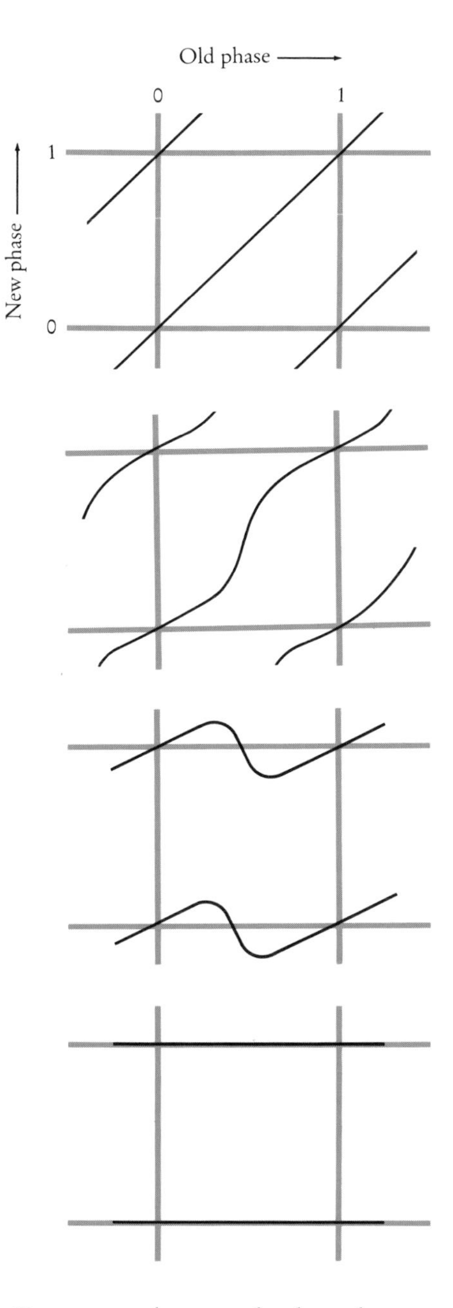

From top to bottom, the dependence of new phase on old phase progresses from odd to even with increase of stimulus size.

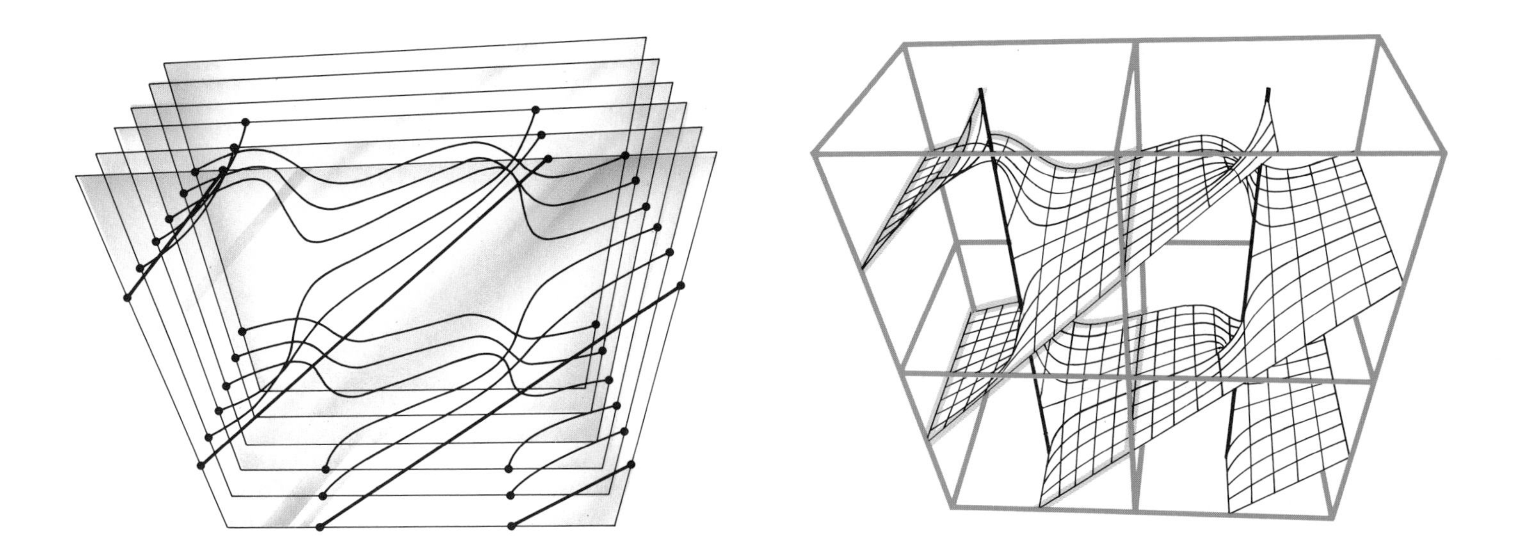

cell, decreasing the stimulus back to zero at the same phase as before (only one cycle earlier). However much we rose or fell going back along the right wall, we have now undone it exactly. So we end up at a point exactly one cycle above our starting place. We have traced out a helix that is the boundary of a surface inside a unit cell of the time crystal. What kind of surface has a helical boundary? Near the boundary it is a screw surface like the spiral pavement in a parking garage. As in the garage, something queer must happen in the interior of any such surface: either the surface must end along an interior hole or discontinuity where phase is simply undefined, or else it must contain a screw axis, a place of infinite steepness. There is no way to heal the hole with a piece of surface that is merely very steep: a screw axis (strict discontinuity) is inevitable. To finally answer a question posed above, the transition from odd resetting (foreground planes) to even resetting (background planes) is made at that screw axis. There the surface must turn vertical or simply vanish; in either case, its altitude is indeterminate. That particular stimulus size, administered at the vulnerable phase, is our singularity, a discontinuity of peculiar violence. Arbitrarily near this ambiguous point, every new phase is accessible by arbitrarily slight adjustments of stimulus size and timing.

Left: Graphs of resetting data can be stacked up from front to back in order of size of the stimulus used. *Right:* The resetting curves combine to define a resetting surface, which shows how new phase depends simultaneously on the old phase and the size of the stimulus. The resetting surface encounters the boundaries of a unit cell along a helical path. The vertical walls (left, middle, right) are at old phase = 0. The horizontal floors (bottom, middle, top) are at new phase = ½.

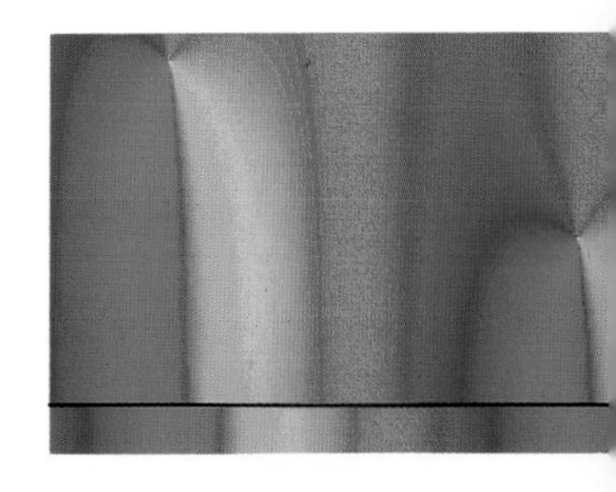

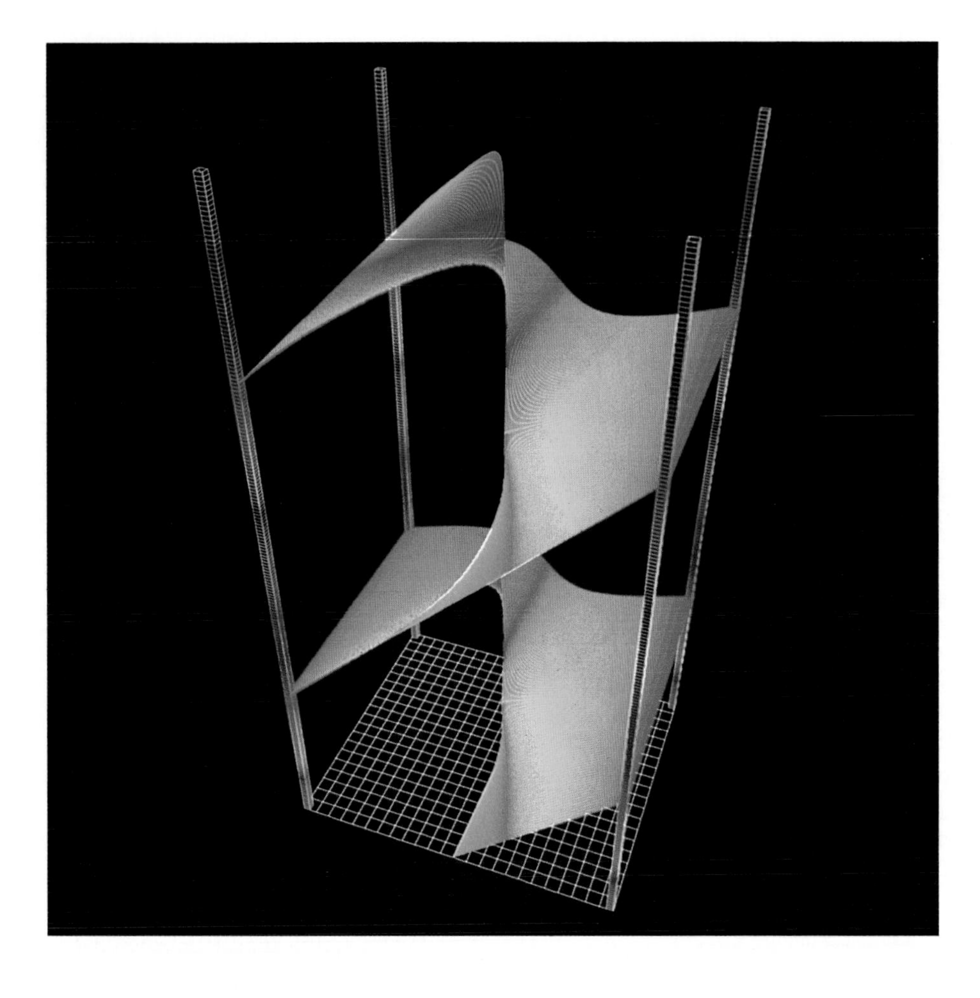

The resetting surface on page 79 color coded for new phase.

Is this really what we foresaw in the colored singularity trap? To find out, let's just apply the standard color coding for new phase. We roll the color wheel up the vertical new phase axis, painting as we go. Imagine colored light from the axis suffusing the time crystal, radiating in horizontal planes from each point of the axis. Where the light strikes the screw surface, it is colored.

Now look down upon the time crystal from above. You see a column of unit cells as a rectangle: old phase spanning a cycle left to right, and stimulus size ranging from zero at the front to maximum at the back. Within the rectangle is a lovely pattern of bright color: the illuminated surface. As before, we can make a tour of inspection around the borders

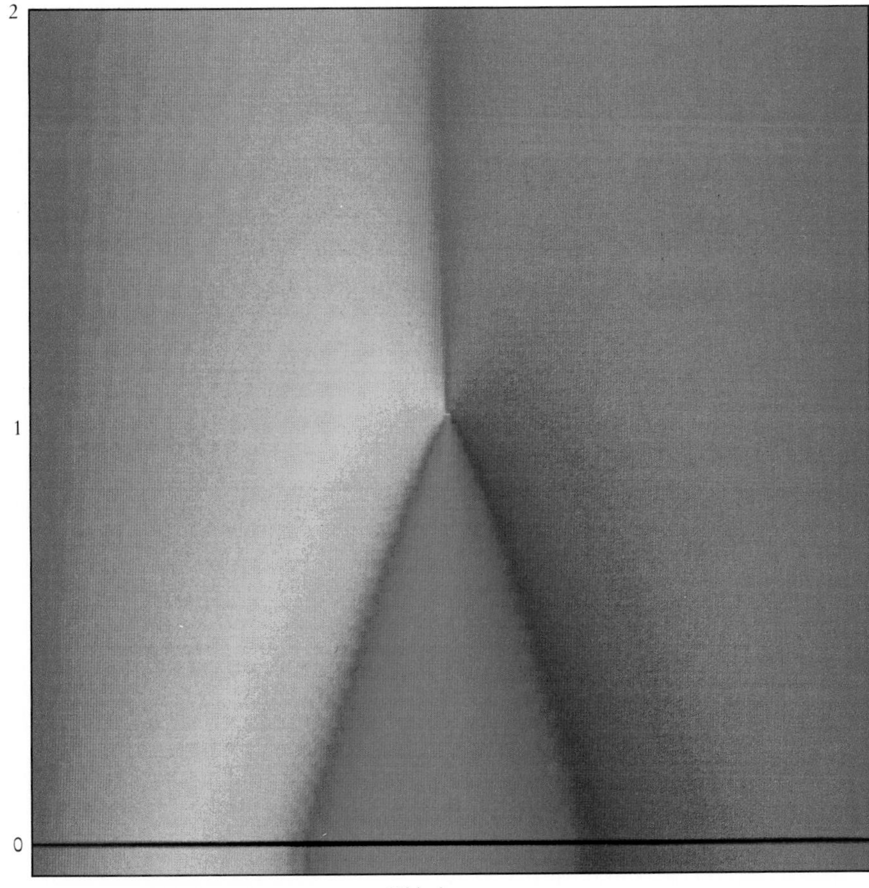

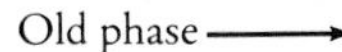

Viewing a unit cell of a resetting surface from above, with new phase (surface altitude) encoded as color, one sees a color contour map similar to the color plot on page 74.

of that rectangle, counting full cycles of color traversed. The count is easy: one cycle across the front, where new phase = old phase; then some change from front to back, which will be reversed in a moment; across the back, no net change, as new phase merely increases and decreases back again in the even resetting pattern; then from back to front reversing the change from front to back, at the same phase a cycle away. The net result? One full cycle of increase. As before, this implies that all the color contours—the time zones—must end somewhere inside. Wherever that happens, somewhere inside the rectangle (and thus somewhere on the surface), there all the hues come together in a hueless point of gray: a phase singularity.

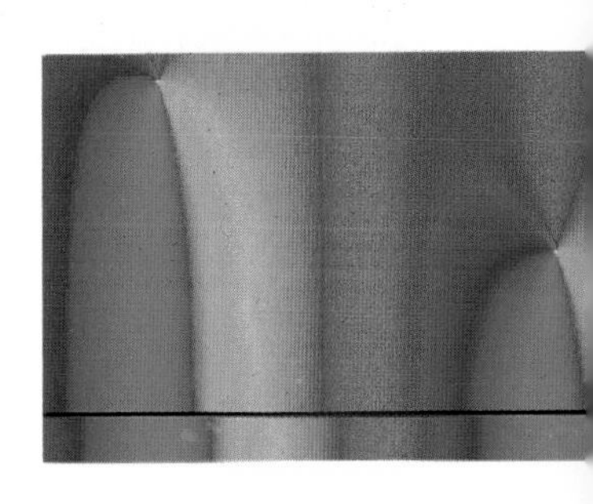

This three-dimensional graph summarizes thousands of experimental measurements of fruit fly eclosion times (see vertical phase rulers) after an exposure to light (duration doubling seven times from foreground to background) starting anytime during the first three cycles of free-run in darkness (see horizontal phase ruler).

Chapter Five

Crystals of Living Time

Quite often it matters little what your guess is; but it always matters a lot how you test your guess.

GEORGE POLYA, *HOW TO SOLVE IT*

The idea that we mammals have internal clocks and that their synchronization is continually renewed by the day/night cycle came from the experiments of Maynard Johnson at Harvard just before the outbreak of World War II.[1] They were promptly forgotten. But Frank Brown at Northwestern University and Colin Pittendrigh at Princeton[2] revived the clock metaphor in the 1950s, and people began to think about entrainment again. The idea of using a daily phase-resetting stimulus experimentally to entrain a circadian clock originated in 1956 with Kenneth Rawson's doctoral thesis at Harvard.[3] This aptitude for resetting on cue is the most fundamental requirement of any circadian rhythm, if it is to be useful in a periodic environment. Rawson's experiments used mammals, though they were inconvenient vehicles for efficient inquiry about cued resetting. Biologists soon sought other organisms, sometimes exotic, each with some special advantage for interrogation.

Experiments at Harvard using *Gonyaulax* were among the first, though it was nowhere yet recognized, to detect even resetting. At the same time or a little earlier, even resetting data had been obtained from the common fruit fly, *Drosophila pseudoobscura,* in the laboratories of Colin Pittendrigh and Victor Bruce at Princeton.[4] A decade later, as a student at Johns Hopkins University, I noticed the even pattern in the data published on *Gonyaulax, Drosophila,* and, by then, several other organisms. Its apparent implication—the phase singularity—was conceptually disturbing and physiologically intriguing. What would be the best experimental organism for the first decisive test of this line of infer-

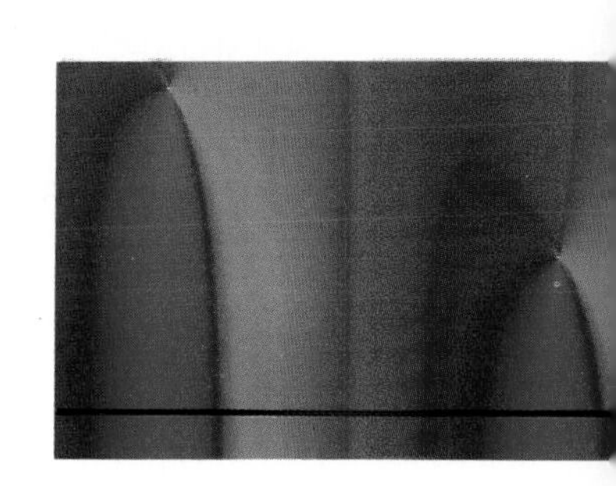

ence? After almost a year of deliberation, I chose the fruit fly and moved to Princeton.

Martin Luther thought flies were created by the devil, since they had no possible practical use, but biologists have since found marvelous uses for them. Soon after geneticists adopted it in the early 1900s, *Drosophila* became the most studied multicellular organism on the planet. Its present huge library of available mutants provides the experimentalist a tool for almost any imaginable trick. Such tools are needed: in any critical experiment, there are a lot of practical details and "control experiments" that must be carefully managed lest the final outcome be ambiguous. Unless the experimental biologist frames his question with exquisite care, Nature will generally find a Delphic reply and the original question will remain unanswered. By using genetic markers to create uncommonly uniform populations for measurement purposes, by studying mutants with various defects of the visual system, by mutating to select for strains with clocks of altered period or temperature sensitivity, and so on, biologists have at this writing gone so far as to isolate, clone, and sequence genes involved in the development of circadian clocks.[5]

The Fruit Fly's Debut

The fly is conceived in the normal way, by a father's sperm cell activating a mother's egg cell. The fertilized egg develops and a tiny larva

Left: A fruit fly *Drosophila melanogaster*. *Right*: Inside the ostensibly dormant pupa of the fruit fly, wholesale demolition of larval organs is almost complete, while the organs of the sexual adult grow and link together. A circadian clock in the brain keeps time undisturbed all the while, and finally dictates the moment of emergence. These individuals of *Drosophila melanogaster* differ in age by a few days, the paler ones being the youngest; the compound eyes are already red in all, just under the transparent "trap door" from which the finished fly will emerge.

(maggot) hatches out. There comes a stage in the life cycle of any fly when it must be transformed from larva to the sexually mature winged adult. During this interval of metamorphosis the animal secures itself in a hard brown capsule like a rice grain, called the *pupal case.* While thus enveloped in a self-contained life-support system, the motionless pupa is very easy to care for. It wants no food or water and extrudes no excreta. Meanwhile almost all its larval organs are dissolving to form a nutritious soup in which the organs of the mature adult grow quickly. But the brain remains intact, and in the brain the circadian clock keeps time. When things have gone far enough, it is time for the virgin female to emerge (for technical reasons, it is convenient to use a mutant strain that seldom produces males), test out her new wings, and find a mate. She inflates a little balloon on her head, forcing open a trap door in the pupal case. Within a few minutes, she's out. In nature, and in a laboratory that is exposed to a 24-hour cycle of equal days and nights, the emergence event, called *eclosion,* occurs in the first hours of daylight.

The moment of eclosion is dictated by the circadian clock, perhaps in the general way discussed at the end of Chapter 2. The pupae grow daily more mature, more ready to hatch out and beat the competition for mates. With readiness rising toward some threshold day by day, another factor must be considered: some times of day are presumably better than others. One might speculate, for example, that it is useless to emerge after the end of daylight, and fatal to emerge with delicate wet skin in the afternoon dry heat. One would like to emerge as early as possible in the morning. An internal rhythm can help by making the threshold for hatching sometimes higher, sometimes lower.

If the threshold varies little, then flies still emerge at any and every time in the circadian cycle, with some times only a little more likely than others. But if the threshold varies widely, at some hours rising more quickly than readiness rises, then no one will emerge during those hours. Depending on an immature pupa's age, she will emerge in one bunch or another, but not at any in-between hour; events will be discretely bunched day after day. The timing of these bunches reveals the timing of whatever clock is expressed in the threshold rhythm.

Embryonic fruit flies do not automatically have functioning circadian clocks. Individuals carefully reared from the moment of conception in constant darkness at constant temperature give no evidence of preferring one hour over another to emerge, except in respect to age. In a population whose age spans several days, individuals will then complete

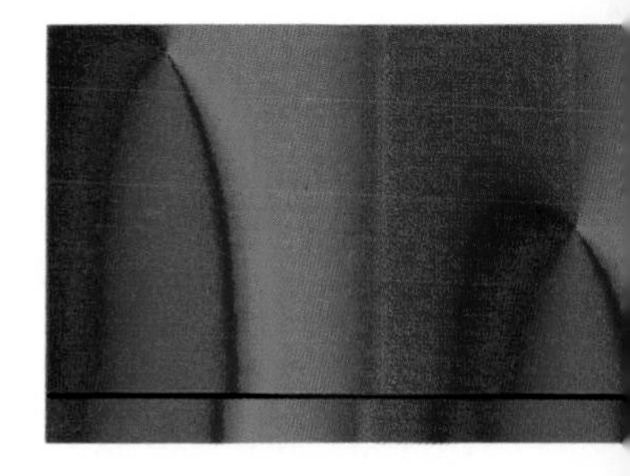

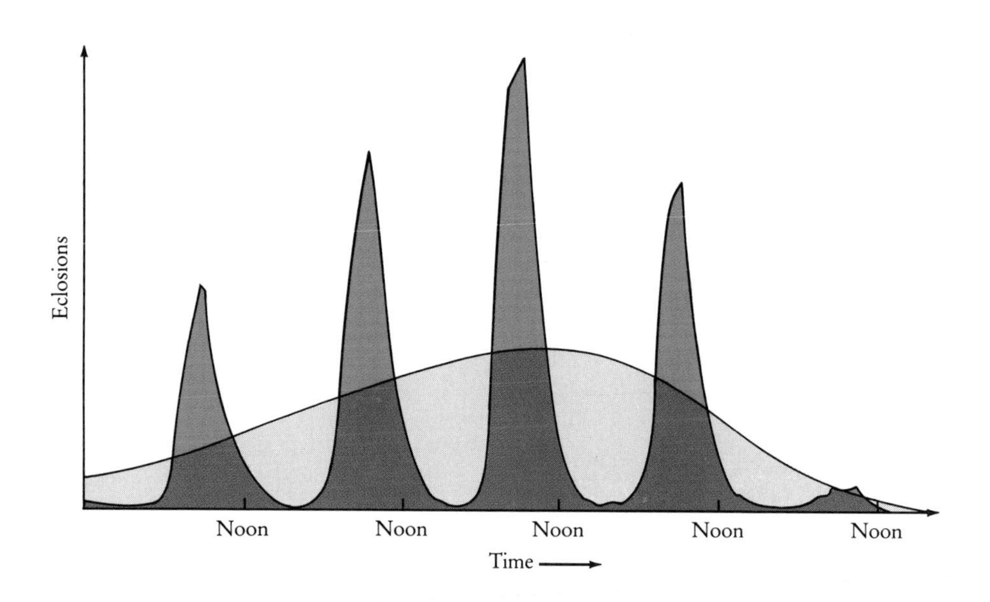

Five successive daily bursts of emergence (colored curve) from a population of *Drosophila* pupae whose age spans about five days. The emergences bunch into discrete peaks 24 hours apart, although the distribution of times at which individuals reach maturity (gray curve) is quite continuous.

their development and emerge in a smooth distribution spanning several days, without daily bunching. But if anything at all is done to give a time cue during early development, then emergence some days later will instead be distinctly bunched in peaks 24 hours apart. Exposure to a single photoflash, for example, starts all the embryonic clocks. (The period of 24 hours is innate and genetically determined; mutants exist whose innate period is, for example, 19 hours.)

Under this arrangement of rearing conditions, the state of the inner clock depends delicately on slight vicissitudes of temperature and darkness. A more convenient procedure entails rearing the larvae under constant light, which suppresses the cycling of their clocks, even if they have already been started. It turns out that fruit fly clocks are blind to red and even yellow light, but even as little blue light as there is in unobstructed moonlight swiftly suppresses their circadian oscillation. Bright fluorescent light serves the same purpose in the laboratory. If pupae are harvested in the light then taken into a red or yellow "darkroom," the clocks are all simultaneously released and begin to mark time in synchrony. A burst of eclosion occurs after about 17 hours, another 24 hours later, another 24 hours after that, and so on until even the youngest individuals have matured and emerged. However, any brief exposure to light punctuating the otherwise eternal darkness will offset the timing of all emergence peaks to follow, as though the light

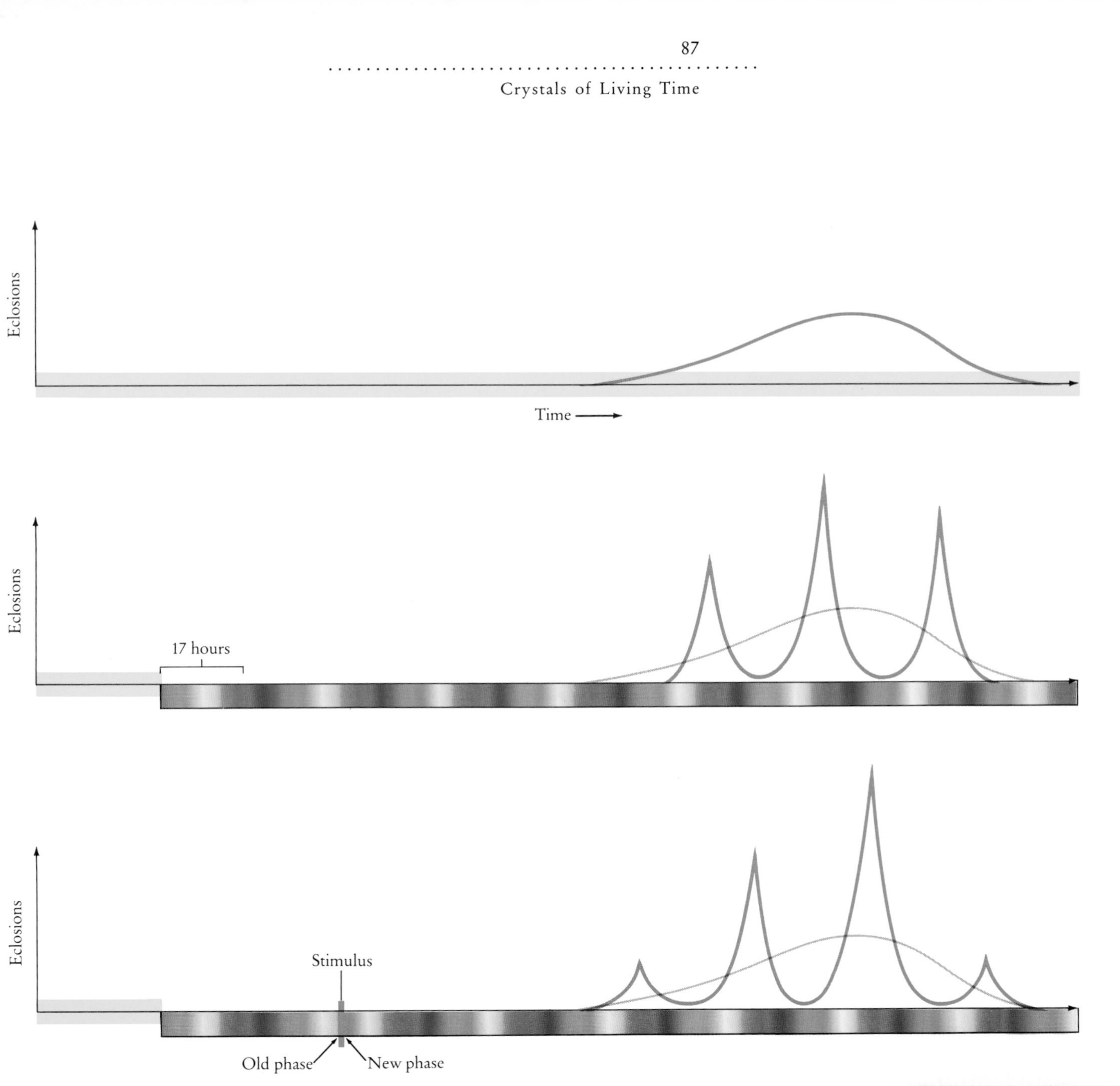

Top: The eclosion curve for a population of flies reared in constant light corresponds to the age distribution of flies in the population. *Middle*: If a population in early pupal stage is transferred from constant light to constant darkness, its eclosions are bunched into periodic peaks. *Bottom*: If a population made rhythmic by transfer to darkness is treated to a later light stimulus, its eclosion peaks may be shifted.

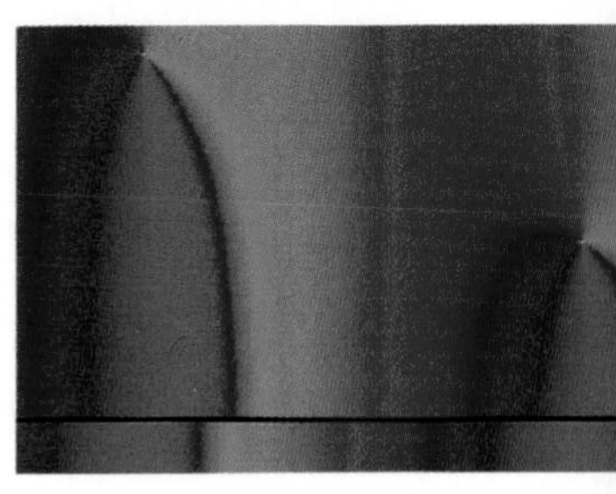

had reset the internal clock from whatever its old phase was at the moment of exposure, to some new phase. The new phase can be inferred directly by extrapolating the subsequent eclosion rhythm back to the moment of exposure.

A Time Machine and Its Product

In experiments on fruit flies, phase resetting is inflicted early in the pupal stage. With machinery that automatically records emergence of flies for weeks, it is possible to observe when the circadian clock strikes "zero"—its phase at emergence—and so to infer its phase at any other moment, since the clock has a period of 24 hours. The biologist can thus know how the fly's clock was reset, say by a light pulse, several days before emergence.

If you watched such an experiment in the back room of my old laboratory at Purdue University, here is what you would see. There are no windows; the floor, ceiling, and walls are painted flat black. Brass and plexiglass devices reflect the bright yellow glare of a sodium safelight, invisible to the little patch of brain that contains the clock. Inside one particular device, a dull black machine about a meter in diameter with twelvefold symmetry, what looks like a tiny crystal bowl of rice rests on black velvet. The bowl's hood is held open by an electromagnet, its contents targeted by a ruby-red beam of light (also invisible to the clock). The "crystal bowl" is a tissue-culture dish and the "rice grains" are pupae lightly glued to the bowl's floor; another eleven bowls rest nearby around a big circle, each closed in complete darkness. In the center of the circle a tilted silvered disk hovers over a hole. There is a

Left: The first step in the experiment to determine the phase singularity of the biological clock in fruit fly pupae: Larvae grow in enriched mashed potatos under continuous lighting. When mature they seek a dry place to pupate, here provided by crumpled plastic film. Afterward, the pupae will be lightly glued to a small dish and set into red safelight. *Center:* Exposed in the time machine to blue light, the internal clocks of the pupae are reset in a way that will only be revealed when they eclose. *Right:* To determine the time of hatching for each group of pupae, the experimenter mounts the 24 time-shifted populations and two controls in the oval spaces roofing an array of 26 teflon-lined red lucite funnels. A sheet of lucite advances once each hour under the array of funnels, positioning the next column of holes milled in its surface. Hatching in darkness, the flies with their still uninflated wings lose their footing on teflon and tumble into the holes, where they are trapped and later counted.

click. A beam of heat-filtered monochromatic blue light shoots vertically up through the hole and reflects off the mirror to irradiate the targeted bowl. The pupae lie motionless, one side of each sparkling in the blue light, the other side still red in shadow. A second passes, or even hours, depending on the programmed protocol, while the animals' brains absorb the light and they are projected ahead or thrown back in their circadian cycles. Then a click, the light source goes dark again, and the mirror rotates in anticipation, to deflect the targeting beam onto the next dish. Sometime later, this time machine comes to life again to open and irradiate that dish, by that time at a later old phase, and send its occupants to a different time zone.

How does this displacement depend on the timing of the stimulus and the size of the kick delivered in blue photons? Only three quantities are involved: the old phase, the stimulus size, and the new phase. First, the phase at the moment of transfer from white light to safelight is a known standard. Adding to it whatever fraction of 24 hours elapses from transfer until stimulus gives the old phase when the stimulus is administered. As a check, the old phase is also inferred from the timing of emergence peaks many days later in a thirteenth dish, a control group never exposed after the initial transfer: that rhythm, extrapolated back to the moment of exposure, also gives the old phase. Any population (and there are a few) in which those two estimates differ by more than an hour or two is rejected: something unplanned and unknown must have happened to some or all of the pupae during that 10-day experiment. Otherwise, the new phase when the stimulus ends (or at intervals of 24 hours after) is measured directly from the subsequent daily emergence peaks: as in the graph on page 87 that rhythm extrapolated back through the moment the exposure ends, or to any moment $24n$ hours afterwards, gives the new phase.

In the first search for the phase singularity of a biological clock, more than 500 dishes were exposed in this way, at various times in the cycle and for various durations, following the singularity trap protocol. For each, the timing of eclosion was monitored to reveal the new, reset phase of the governing circadian rhythm. From each dish there were several bursts of emergence, one each day at the same internal clock time. The pupae who were already old enough are the ones who finished in the first eclosion peak. Their younger sisters needed more days to mature; their emergence was governed by the same circadian rhythm on the later day when they were finally watching the clock in readiness to start life as a sexual adult. Altogether about 1500 bursts of emergence

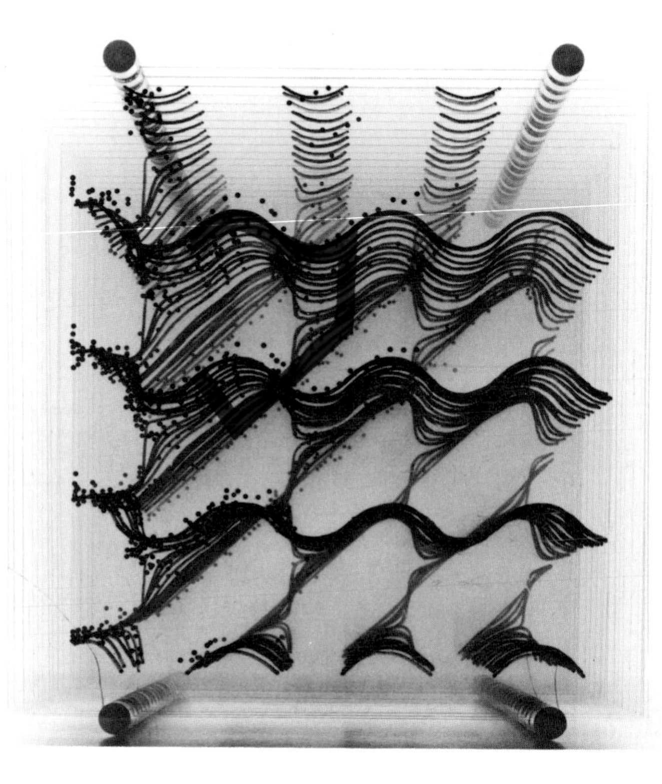

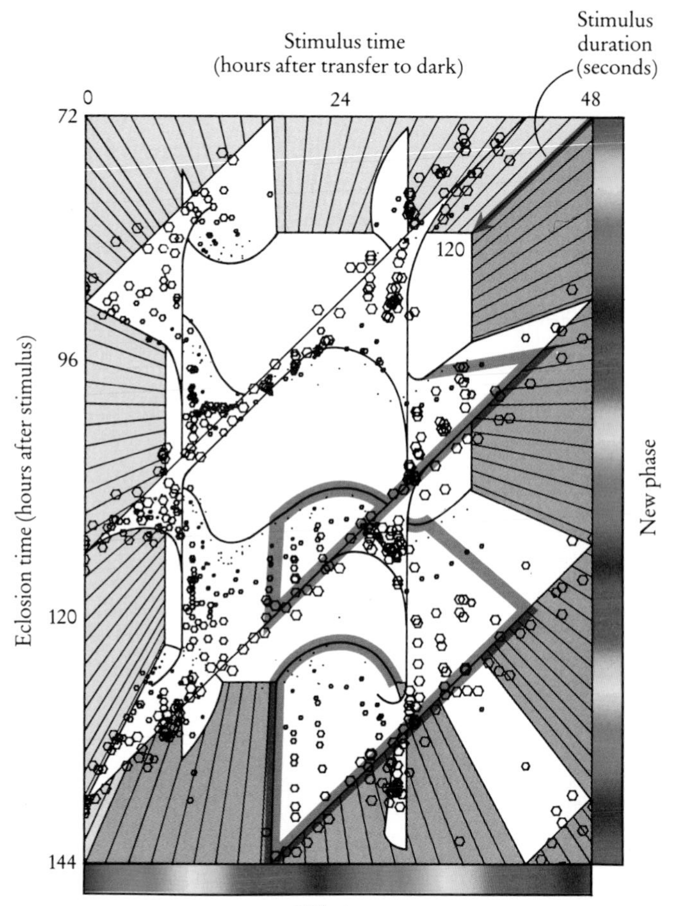

Left: The time crystal of *Drosophila pseudoobscura,* with mathematically defined "ideal" curves and enough data to span almost four cycles in each direction, was first plotted three dimensionally in 1971 in the Medical Research Council's Laboratory of Molecular Biology. In this photograph of the original perspex model, old phase increases to the right, stimulus duration increases from background to foreground, and new-phase measurements increase upward. Six unit cells from the upper left corner are repeated in the adjacent computer-plotted version; there, old-phase and new-phase coordinates increase to the right and upward as here, but stimulus duration increases into the background.
Right: Drosophila's time crystal, measured with pulses of blue light. Transfer of pupae from constant light to complete darkness always starts their clocks at the same phase, so time after the transfer is equivalent to the old phase of the clock when the stimulus is applied. Time from the stimulus to subsequent eclosions is measured downward, spanning three cycles of 24 hours starting three cycles after the stimulus (little happens until then). Data points represent emergence peaks, indicating by definition the moment in time when the clock phase reaches zero. The inferred new phase of the clock when the stimulus ended can be read from the vertical phase ruler at the right. The helical edge of a pair of unit cells of the crystal is outlined in color.

were plotted, each as one dot in a time crystal, a complete three-dimensional graph of new phase in its dependence on stimulus size and old phase. In the complete time crystal, the old-phase axis must extend horizontally for several cycles, because some pupae waited days before the time machine turned its beam to their dish. The new-phase axis must extend several cycles vertically because every day more matured pupae hatched. The crystal shown on the left on the facing page has almost four cycles in each direction. Following the plan of the figure on page 79, it was actually built of perspex sheets in a Cambridge workshop where x-ray crystallographers had built such stacks to visualize the molecular crystals of protein molecules (before it became more convenient to use three-dimensional digital graphics systems). It is here shown with the big-stimulus resetting curves (even type) in the foreground. The helical border of one unit cell is outlined in red. With the advent of computer graphics, it has become possible to dispense with artificially rigid planes. The figure on the right presents the same data from a different perspective, overlaid by a hand-sketched screw surface. Small-stimulus resetting curves (odd type) are presented in the foreground, covering only two cycles to the right and three cycles vertically (879 bursts of emergence).

What kind of surface do these dots outline? The foreground data are plotted as big hexagons and those at the rear are mere dots, and the helical border of one unit cell is outlined in red. Follow the "new phase = old phase" diagonal upward in the foreground, using a zero-duration "stimulus" given at later and later old phases. This is the odd pattern of resetting. Follow it upward to the right across one cycle in both directions ending at the right in a refractory phase, when no amount of light bothers the clock: if you now leave the diagonal to follow the data toward the rear, you see them hover at constant altitude as the stimulus increases to maximum. (There is no logical necessity for such a refractory phase to exist, and in other situations it doesn't, but constancy along the sidewalls plays no essential role in the interpretation.) Then at maximum stimulus, follow the data back to the left across a cycle of old phase: see them ride up and down on the horizontally wavy surface without progressing vertically through any net distance of new phase. This is the even pattern of resetting. Here at the back of the time crystal, exactly one cycle before the refractory phase in which you increased the stimulus, we find the same refractory phase. Now decrease the stimulus duration, holding old phase fixed: as you come forward, it makes no difference to the new phase. And you end not quite where you started,

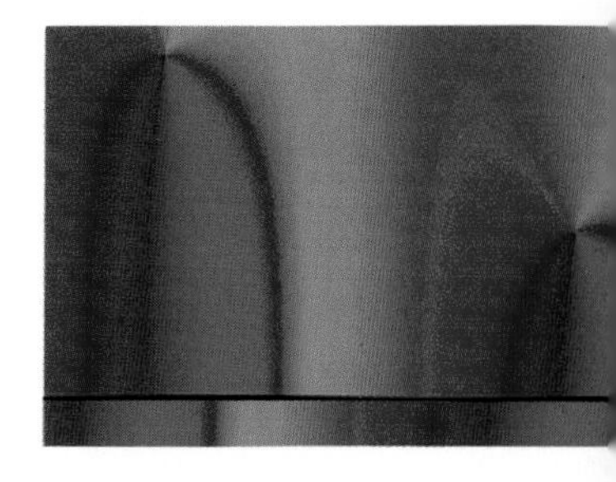

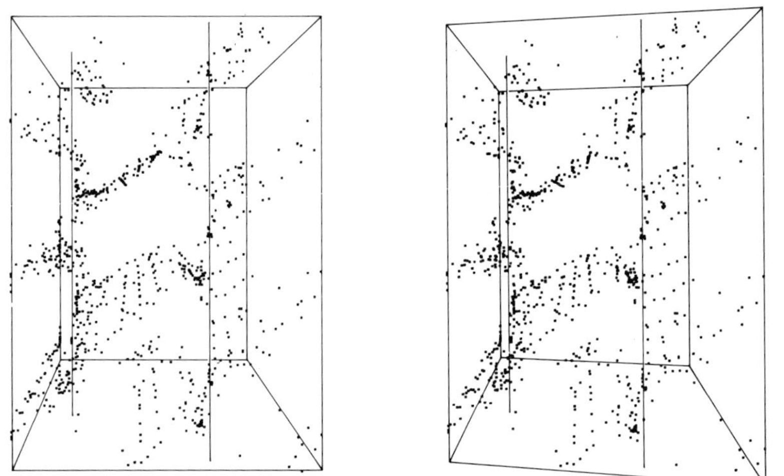

The data of the *Drosophila* time crystal in stereo. The two interior vertical lines depict screw axes of the fitting surface; the second is more in the foreground because of dark-adaptation of the pupae. To see it that way, place an at least 5 inch high piece of cardboard edgewise between the two images. Bring your forehead down to the cardboard and look at both images. After a moment, the two images should fuse. You may also view the images through a stereopticon, one of which was supplied with an earlier volume in this series, *Perception* by Irvin Rock.

but one cycle higher. This circuit embraces one complete unit cell of the time crystal in one turn of a helix, so the surface defined by the cloud of data points is a screw surface, joining smoothly along the red helix onto adjacent screws.

This is a remarkable situation. The tour of inspection around the fringes of a unit cell used only experiments that did nothing, plus a series with such a strong stimulus that the rhythm is reset to practically the same time in every case. One might not think that such a superficial inspection of clock behavior, with such unexciting results, could contain the seeds of something rather peculiar. But the biological clock, as G. K. Chesterton once remarked in a different context, "looks just a little more mathematical and regular than it really is; its exactitude is obvious but its inexactitude is hidden; its wildness lies in wait."[6] The inescapable inference (which is why the experiment was done) is that any helical boundary must support a screw surface, and a screw surface must have an axis inside, somewhere far from the stimuli used to feel out the smooth boundary. The axis of each screw is a vertical line along which the altitude can only become indeterminate. That line lies above a point on the floor. The point is a stimulus, a particular combination of duration and timing of exposure to light. What happens when that particular stimulus is given?

First of all, what is that stimulus? If the blue light is barely bright enough to read by, it is about a minute's exposure. It must be applied close to midnight of the usual circadian cycle, a time when in nature one sees little light. The most a wild fruit fly might see at that hour is moonlight; a few minutes' full exposure would constitute a singular stimulus, but the little pupa, hidden under leaves or buried in the sand, has little chance to see it. Even if it were so exposed during the vulnerable phase, exposure to full sunlight would soon follow, completely overriding the singular effect.

But in a twentieth-century laboratory, organisms encounter conditions never experienced in nature. Their circadian clocks run freely without cueing, naked to probing stimuli and to our curiosity about the clocks' innate responses. What happens in the little dishes of "rice" after the singular stimulus is found and administered? It turns out that the subsequent timing of circadian behavior is indeed indeterminate, just as the helical pattern of rhythm resetting foretold. But there are several alternative ways in which circadian timing could be ambiguous. Did the pupae die, never hatching at all? Or did they continue to hatch in daily pulses, just at a daily hour that no one could guess beforehand? They did neither: they simply hatched all the time without interruption. After the singular stimulus, until such time as some other little stimulus gives the dormant flies a cue from which to mark time, they remain timeless, much as though their embryonic clocks had never been nudged off equilibrium in the first place. Either way—never cued since conception, or cued to rhythmicity then put back into suspended animation by the singular stimulus—the circadian clock lies inert in each fly. In this state, individuals wake from metamorphosis at all hours, but not in rhythmic daily bursts. Moreover, if circadian timing of eclosion is restored by a second light pulse, the new phase is independent of the time when that stimulus is given: it is the same phase as results if no time at all intervenes between the first and second stimuli, i.e., as if the first stimulus were of combined size. After receiving the first dose, the fly's circadian system languished in suspended animation until more light arrived.

The measured resetting surface is like a wonderful snowy mountain with endless descents possible by skiing helically down any of the spiral slopes that surround the singularities. Viewed thus as a hillside, it must have a contour map. That map is shown here for three consecutive unit cells adjacent in the old phase direction, color-coded for new phase as foreseen in Chapter 4.

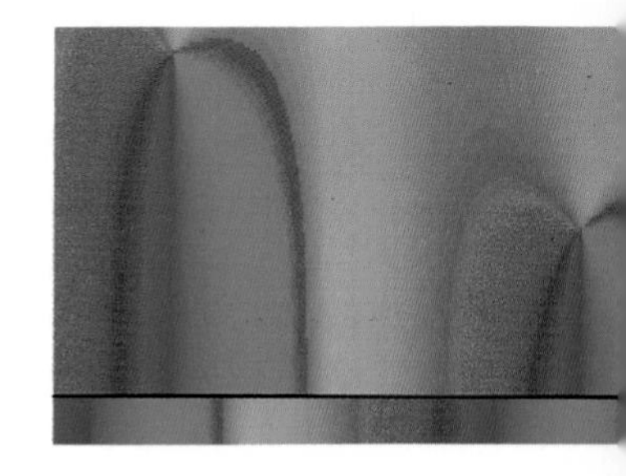

Besides showing the exact coordinates of the singularity and its daily repeats, the contour map reveals an unforeseen squeezing of all contours toward the foreground (small-stimulus side). This shows that later after the transfer from constant bright white light to darkness, a given

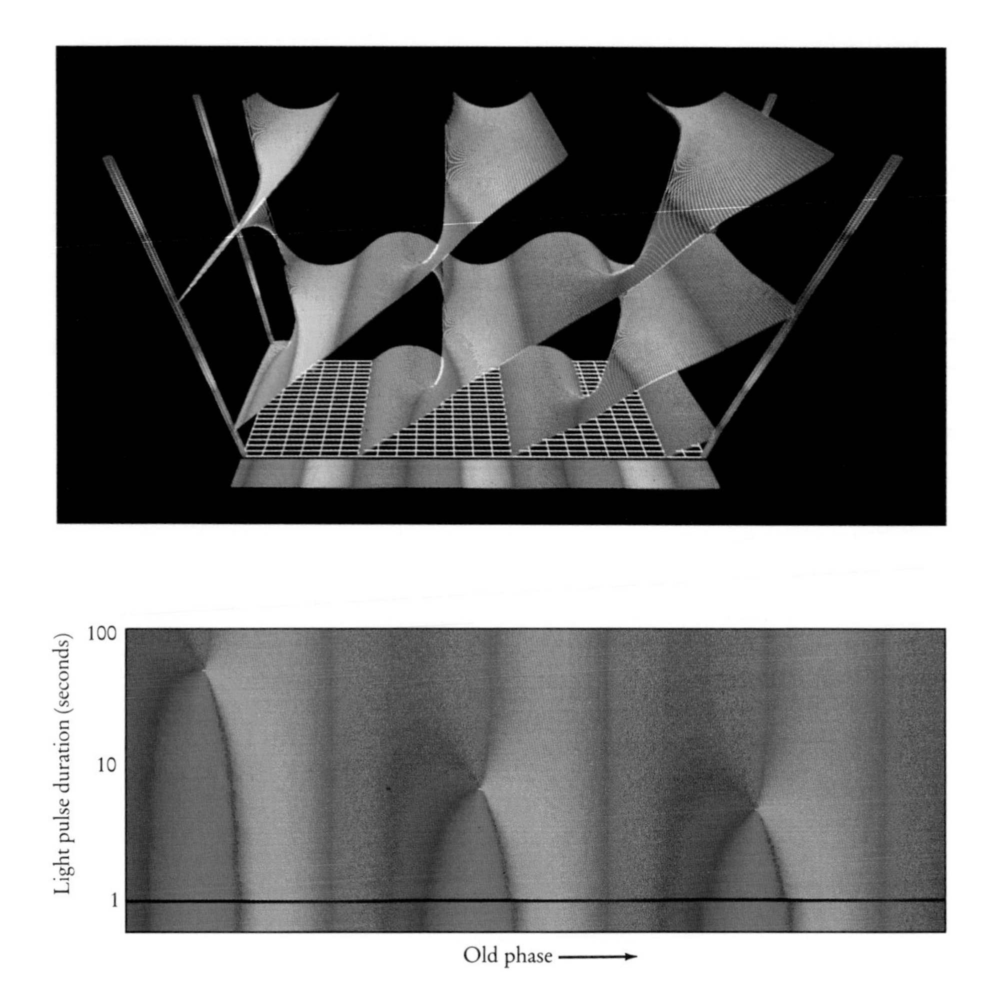

Above: *Drosophila*'s resetting surface across three cycles of old phase (horizontal) and two cycles of new phase (vertical). The colors represent altitude, equivalent to new phase. *Below*: In a view from directly above, the resetting surface appears as a colored contour map of new phase. Exposure duration increases linearly along the stimulus axis of the time crystals on pages 79, 80, and 90, but here it is increased exponentially from 1 to 100 seconds to compress the contours into a shallow rectangle despite the twentyfold change of sensitivity during the first three days of darkness. Because the bottom duration is not quite zero, bottom coloring differs from the phase ruler.

effect is obtained with shorter exposure to the stimulus light. In other words, the fly's clock gets progressively more sensitive to the light, the longer it remains in darkness. This twentyfold dark adaptation runs its course in two or three days: soon after removing the pupae from continuous bright light, exposures are perceived relatively faintly, but a few days after plunging them into darkness, the clock has become so sensitive that continuous illumination dimmer than moonlight suffices to arrest it. This kind of asymmetry about the time crystal has since proved to be common, but no one knows the cause: it could represent regeneration of a pigment that was bleached away during prolonged bright light, or it could represent a slow winding down of the circadian clock, so that the attenuated rhythm is more easily upset—or both. The balance of evidence is on one side in some species, and on the other in others; mechanisms probably differ.

On a spiral staircase the steps protrude radially near the "singularity"; each step edge is a contour line of altitude.

A Pinwheel Experiment

If the unit cell of a screw surface is seen as a spiral staircase, the singularity stands out as a convergence of the edges of the steps toward the supporting pole. Thus, the contour map of the staircase resembles a snapshot of the earth from high above the pole, with all the time zones converging to their singularity. But the time zone or new phase that a clock lands in tells what time the clock shows only momentarily; its hands continue to turn. In the same way, a phase contour map only shows the phase of the circadian clock at a particular moment, a whole number of cycles after the stimulus. Thereafter, the phase changes. In other words, the locus of a given phase is a wave proceeding from one time zone to the next like the dawn on a polar map of the earth, spinning like one arm of a pinwheel. This moving locus of uniform time has a name—*isochron*. The original design of the experiments to detect phase singularities called for measuring the isochrons and monitoring their movement. The measurements at that time were most conveniently done piecemeal, sampling the isochrons at hundreds of points in as many separate single resetting experiments. But ideally, the arrangement would have been as follows:

A lawn of pupae is spread out on a desk top under constant light to keep their circadian clocks arrested (see the figure on the next page). By slowly moving an eclipse of the overhead light bulb from east to west, we permit the circadian rhythm to start in column after column of pupae. If this creeping nightfall takes three days to engulf the whole desk top, an east-west gradient of phase is established, spanning three cycles. With all pupae now steadily oscillating in the dark, they are exposed to light, almost simultaneously (say, within two minutes), by moving the shadow southward, exposing first the north edge of the desk and eventually the whole desk. Just before the shadow is withdrawn from the south edge, it is suddenly moved back up to re-cover the whole desk top. Thus each row of pupae was exposed to a resetting stimulus for a time proportional to its north-south position. Those at

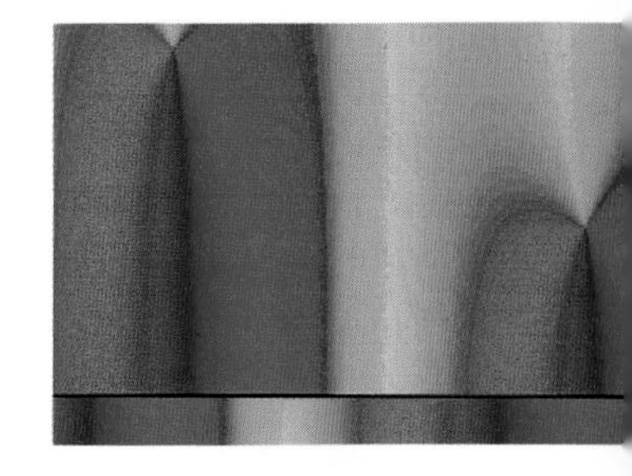

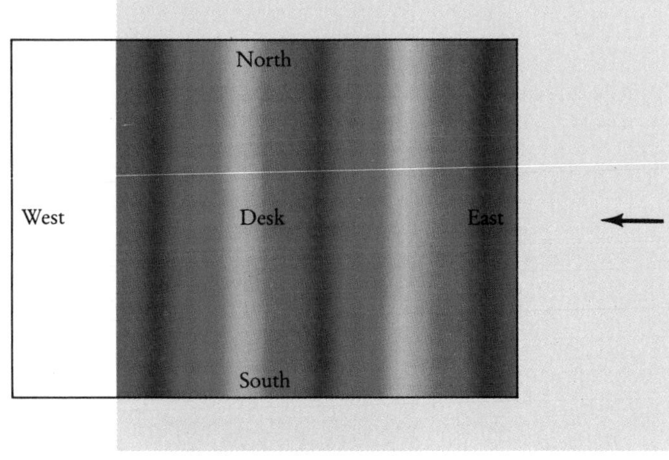

Establishing a gradient of old phase

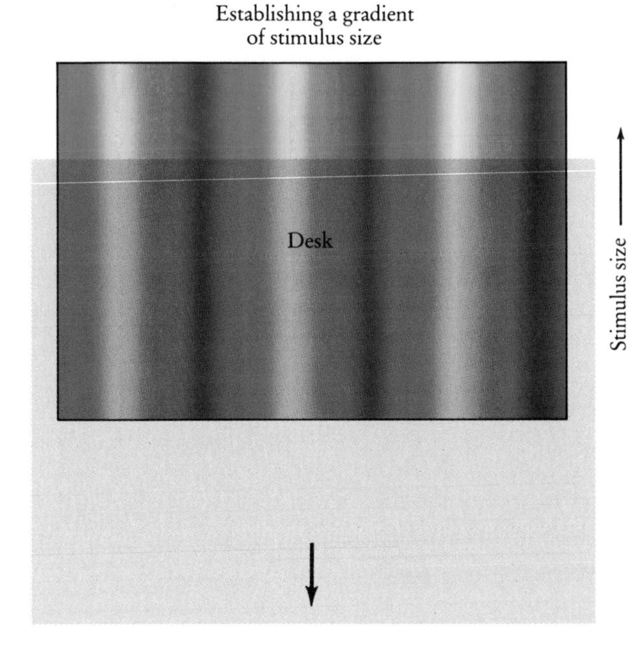

the north edge were exposed for a long time; those at the south, not at all—we have a north-south gradient of stimulus. In short, we have set up on one desk top the whole array of individual phase resetting measurements. The results will be visible as a wave of emergence that shows from one moment to the next which pupae are at eclosion phase in their cycle. As the flip sequence of color-coded phase contour maps shows, that isochronal wave (red) would spin like a pinwheel about a peculiar spot where they hatch continually or not at all—hence the name given in anticipation of its trial, the *pinwheel experiment.*

Although we focused here on the original *Drosophila* experiment, the same protocol has since been followed with several other species of insects,[7] plants,[8] and even unicellular organisms,[9] with similar results. Beyond the narrow domain of circadian rhythms, time crystals and phase singularities have been discovered by the same procedures also in the periodic biochemistry of sugar metabolism,[10] in the half-hour-long water-transport rhythm of oat seedlings,[11] in the rhythmic contractions of the leech's heart[12] and the cat's heart,[13] and even in the nervous control of breathing in the cat,[14] with possible bearing on the mystery of sudden infant death.

A pinwheel experiment. An east-west phase gradient is established on a desk top covered with circadian clocks by moving a shadow as the arrow indicates. The resulting phase gradient is indicated with the standard color code. A north-south stimulus gradient is then established by pulling the shadow to the south then quickly replacing it. We expect to see contours of equal new phase similar to those on page 94. To see the result, flip the lower right page corners from back to front. The color pattern rotates clockwise about a singularity whose vulnerable phase falls 7 hours after darkness began, and another at 31 hours (the flip movie animates the left half of the bottom figure on page 94). This movie was painted by computer from the observed emergence times of some hundreds of thousands of fruit fly pupae.

A VULNERABLE PHASE IN THE HEARTBEAT

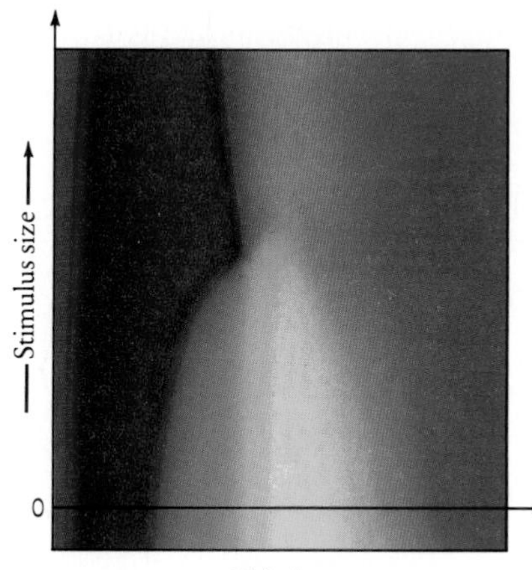

Color-coded new phase calculated from the electrophysiological equations of the pacemaker of the heart. Red represents phase zero, the moment of the heartbeat in a cycle of about one second.

The electrical mechanisms of pacemaker tissues in the nervous system and in the heart are well enough understood to permit the prediction of odd and even resetting and a phase singularity in response to the kind of stimuli that normally control such pacemakers.[15] This theoretical projection has been checked in the laboratory and found to be valid in some cases (and possibly invalid in other cases, when normal membrane processes proceed so abruptly that topological inference becomes irrelevant). Applied to the pacemaker node of the heart, the ephemeral singular stimulus, applied electrically, may induce lasting quiescence; but more commonly, the node merely skips a beat and recovers with unpredictably offset timing. The figure on this page shows the pattern of new phase as a function of old phase and size of an excitatory electrical stimulus applied to the pacemaker node.[16] The whole mass of rhythmically active muscle normally controlled by the heart's pacemaker organ also should be vulnerable to permanent arrhythmia consisting of rotating waves, if exposed to an electrical stimulus of the right size—neither too small nor too large—and if it arrives within a narrowly defined interval of less than one-tenth of a second during the normal heartbeat.[17] This "vulnerable period" has long been known empirically, since artificially induced fibrillation was discovered in 1914. However, it has only recently been confirmed that the disruptive stimulus must also lie between definite upper and lower limits of size, and that the evoked arrhythmia in fact consists of mirror-image counter-rotating waves, as required by theories based on the phase singularity.[18] Is this a coincidence? No one yet knows.

The take-home lesson from any pinwheel experiment is that the isochrons do converge in an orderly way to a point, a phenomenon that has no place in some models of the clock mechanism. In drawing attention to that unexpected point, the experiment still leaves unresolved a fundamental ambiguity. What does it mean to annihilate a biological rhythm? Is the clock merely paralyzed somewhere along its cycle or is it in some state of suspended animation equally remote from all states on the normal cycle? Has every part of the organism lapsed into a stupor of timelessness or has it only lost the former coherence of microscopic oscillations that still continue, randomly reset?

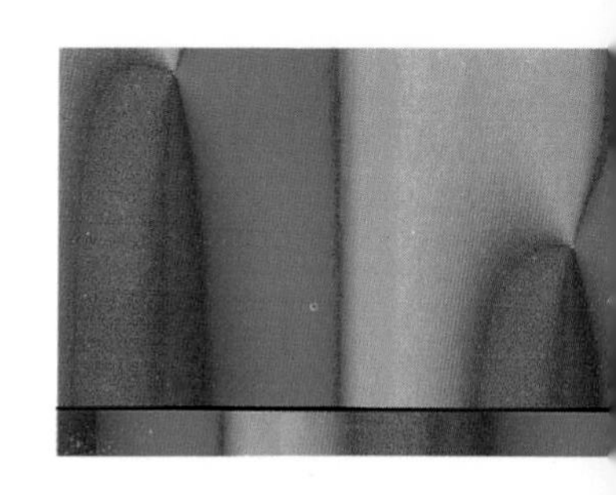

Flowers of the plant *Kalanchoë* open and close spontaneously, each with its own, autonomous circadian rhythm. The rhythm of each flower, moreover, seems to be a collective expression of many cellular clocks.

Chapter Six

Collective Rhythmicity in a Community of Clocks

He tampered with the clocks to see if they would go, out of a strange perversity, praying that they wouldn't. Tinkers and tinkerers and a few wizards who happened by tried to start the clocks with tools or magic words, or by shaking them and cursing, but nothing whirred or ticked. The clocks were dead.

JAMES THURBER, "THE THIRTEEN CLOCKS"

Imagine a clock-shop wall hung with hundreds of mechanical clocks all set to chime at noon. If they are not all set correctly, some chiming will start in late morning, the cacophony will reach a crescendo near noon, and the last chimes will be heard in midafternoon. A blind observer on the street, unable to distinguish individual clocks, might say that the clock shop displays a smooth circadian rhythm of noisiness. If he were statistically inclined, he might fit a smooth sinusoid to the data and describe his observations by reporting the amplitude of that wave and the times of its peak and trough.

Now suppose a horde of gremlins invade the clock shop and briefly hang weights from all the hands in the fashion explored in Chapter 3. Every clock is reset, each according to its old phase. On the next day our street observer will report a change in the phase and the amplitude of the collective rhythm of noisiness emerging from the clock shop. It is not a difficult matter for theorists to write equations for the new phase and amplitude, as functions of the size of the weights and the initial distribution of old phases when the weights were hung. What would you expect? If you remember that the individual clock's resetting curves are all of odd type, and that they never show even resetting or a singu-

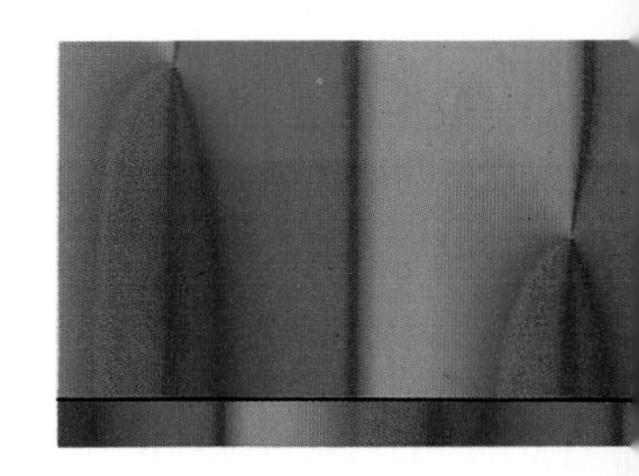

larity, then the answer might surprise you: the new phase of the collective rhythm turns out to depend on its old phase and the stimulus size exactly as plotted in the time crystals that you have seen in this book! The reason, in brief, is that the collective rhythm, unlike an individual clock, has an amplitude related to the synchrony of timing among clocks. Not only phase but synchrony as well is strongly affected by a stimulus. When synchrony is good, the collective amplitude is high and phase of the collective rhythm is clear. When synchrony is poor, amplitude is low and phase is ambiguous. In detail, the population's time crystal turns out to be a screw-shaped surface twined around a singularity.

Resetting experiments, then, are not to be taken at face value until it is established that the rhythm observed comes from a single clock or at least a collection of synchronous clocks. If that cannot be established then not much can be inferred from time crystals and singularities about the nature of the individual clock. The individual clock's works remain invisible as long as it is allowed to participate anonymously in a clock shop.

After a singular stimulus, what does the absence of rhythm signify, for example, in a teeming population of individual flies? Is every organism's personal clock defunct? Or is each perhaps still running normally, just randomly reset relative to any other fly who was subject to the same stimulus? Can a biologist know by looking only at the collective rhythm of a population? As Sherlock Holmes remarked,

> That is a question which has puzzled many an expert, and why? Because there was no reliable test. Now we have the Sherlock Holmes test, and there will no longer be any difficulty.

There is a test, and it provides a decisively clear choice between distinct outcomes: the population's response to a second stimulus should be a clear rhythm, but it should look different in the two cases imagined.

Consider first the limiting extreme case in which every single fly in the population had her clock turned off. That could have happened if their clocks were all exactly synchronous when the stimulus occurred, and if they were all identically exposed to it, and if its timing and strength were just right, and if the individual fly's clock has a stable equilibrium like the cuckoo clock in the box on the facing page. That is a tall order, but anyway, consider the consequence: all the clocks are then at rest under identical conditions, so when a later stimulus simultaneously strikes them all hard enough to restart them, they restart in

This mechanism keeps the pendulum swinging—unless at just the right moment the bob is pushed just hard enough to leave it hanging stably at equilibrium.

OF PENDULA AND CUCKOO CLOCKS

The pendulum that hangs beneath a cuckoo clock is itself a clock of much shorter period, typically about one second. If you gently bump this oscillator to bleed off a little of its momentum or give it a little more, you will have altered the pendulum's amplitude momentarily, while inflicting a permanent offset on its timing. You might have kicked it a quarter-second ahead or behind if you gave it a vigorous tap, or somewhat less if you were more gentle. Anyway, the new phase of that one-second oscillation depends on the old phase when you struck and the vigor of your tap. It turns out that the manner of dependence is the very pinwheel pattern discovered in biological clocks. In particular, you can demonstrate the phase singularity quite simply, as David Paydarfar first showed me in his father's house. Bump the pendulum just when it passes the bottom of its arc, at its moment of maximum speed, not too hard. Bump it just hard enough to neutralize that momentum. It hangs straight down, stably motionless. Little breezes won't start it again, but a big enough tap will; then it quickly builds back up to standard amplitude.[1]

step. The population rhythm, being a sum of identically phased normal rhythms, will look normal and will have a certain characteristic phase that depends on the size of the restarting stimulus, regardless of when it was given. That is the diagnostic for the first case.

Now consider a second limiting case (remembering that reality will probably hover somewhere between these extremes). Suppose the circadian clock can't sit still unless actively suppressed by an inhospitable environment. Arrhythmicity in the dark then can only mean that every fly in the population merely got her clock reset, but at random relative to others in the same population. This interpretation of the time zones and their measured convergence near the singular stimulus might be appropriate if the singularity represents an *unstable* equilibrium: oscillations immediately resume, but with random timing. Apparently random resetting might be expected near any convergence of time zones, since new phase there depends very delicately on the exact timing and

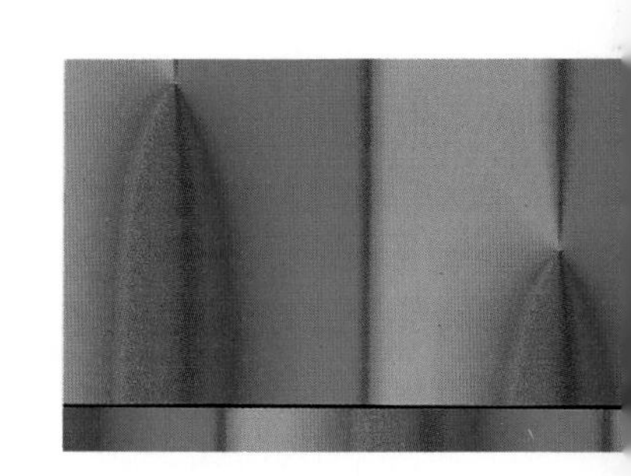

strength of the stimulus—or equivalently, on the individual's exact old phase when the same stimulus was given to the whole population.

If this were the right interpretation of collective arrhythmicity, then it would be easy to anticipate the effect of some later stimulus: the stimulus would reset normally rhythmic flies from each and every time zone to a new time zone as given by the familiar resetting curve for that kind of stimulus. Average those diversely reset rhythms and you would have a new collective rhythm with features somewhat broadened by smearing over the range of new phases represented. The expected sum of individ-

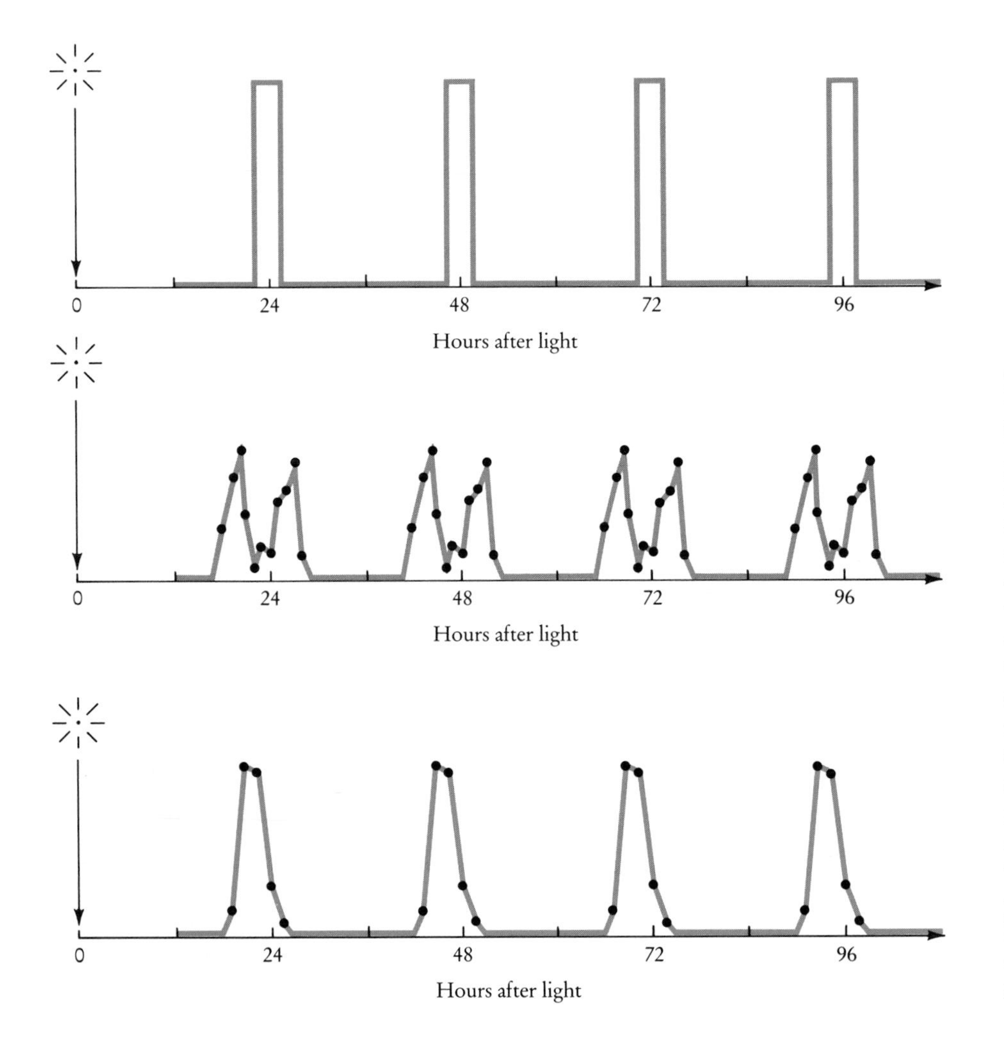

Top: If an arrhythmic population of pupae are all at stable equilibrium, and then their clocks are restarted by a brief exposure to light, this is the expected distribution of their arrivals at phase zero. *Middle:* Here, the supposition is that pupae were uniformly scattered throughout the normal circadian cycle, and were phase reset by the light pulse in the normal way. The expected result can be calculated from the measured resetting curve for that stimulus. Here it is shown instead as the result of an experiment in which the scattered population was deliberately assembled by pouring together diverse populations, each cycling synchronously at different phase. *Bottom:* The result of an experiment using a population of pupae made arrhythmic by a singular stimulus is clearly different from the second possibility and resembles the first.

ual rhythms can be worked out mathematically. Or you can measure the characteristic phase and amplitude of that expected smeared-out collective rhythm physically by pouring together pupae reared in 24 different time zones to create an arrhythmic population of individually rhythmic flies. Either way, the resulting collective rhythm would be the same, and that rhythm's shape is diagnostic of this second limiting case.

Where between these extremes does the real population in fact appear? Almost at the first extreme, in the case of *Drosophila pseudoobscura:* the individual fly's clock is nearly at rest after the singular stimulus. Does this singular state of the fly resemble any familiar condition? Yes. It seems the same as the arrhythmic state of an embryonic fly (young larva) before it has ever experienced light or a change of temperature or any other cue from which to mark time periodically (Chapter 5). The same test applied to pupae so reared gives almost the same results. In making the flies timeless, we seem to have returned their clocks to their developmental origin.[2]

There is another, more straightforward test of the two alternatives—simply monitor the personal clock of each individual before and after the singular stimulus. As it happens, *Drosophila* is an inconvenient subject for this version of the experiment, but a related species cooperates quite nicely.

When Do Mosquitoes Sleep?

If someone tells you a surprising fact about himself, you probably wonder whether you have been favored with a penetrating insight into human nature, or only gleaned a tidbit of gossip about a particular person. So you reserve judgment until you hear others make similar comments. Experimental biologists like ~~to~~ speed up this process deliberately, to find out how widely a new observation holds among related life forms. Experiments can be contrived to ask, Which species reset their clocks in the even mode? What do their time crystals look like? What happens after their clocks are forced to the timeless singular state?

Closely related species were interrogated first, naturally. A mutant fruit fly whose clock had a 19-hour period was tried and found to be little different.[3] But that's almost like asking a twin sister. Eric Peterson, who was then a graduate student in England, sought a more instructive variation, some kind of insect whose circadian rhythm could be followed cycle after cycle in one individual. With this convenience he

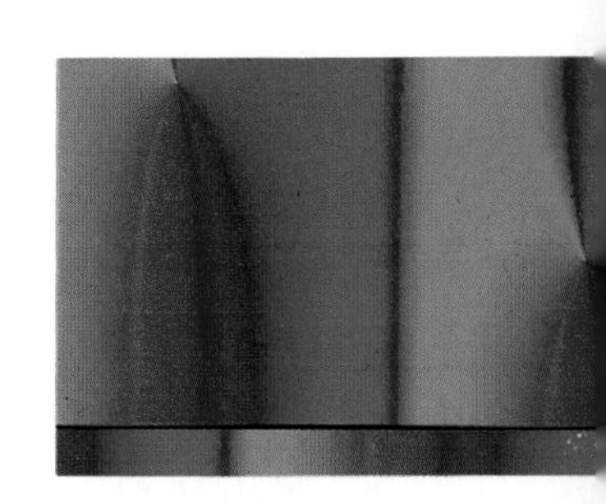

could check the conclusions first reached only by indirect inference from collectives of clocks.[4]

Peterson found a cooperative species in our ageless enemy the mosquito. As we saw in Chapter 3, the *Culex* species he chose is active at dawn and at dusk and rests in between. In the timeless conditions of temporal isolation, this pattern persists: two bursts of activity about 12 hours apart mark each 24-hour cycle. Electronic devices record this rhythmic alternation of sleep and waking in a little cage made to accommodate a solitary mosquito. With lots of cages, Peterson could conduct the successive stages of the singularity-trap procedure efficiently, each cage testing the psychological impact of a different stimulus at a different time in the cycle. The mosquito turns out to have a time crystal that looks a lot like *Drosophila*'s.

Left: The circadian rhythm of a single mosquito in temporal isolation in double-plotted raster format. There are two bursts of activity in each cycle of about 25 hours. Other individual mosquitoes run with periods closer to 24 hours or even less. *Right:* Two individual mosquito insomniacs in the aftermath of the singular stimulus.

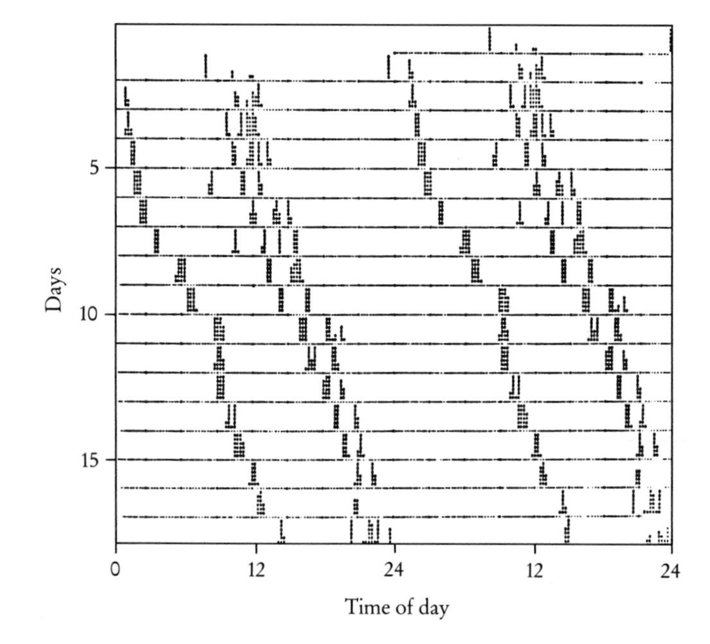

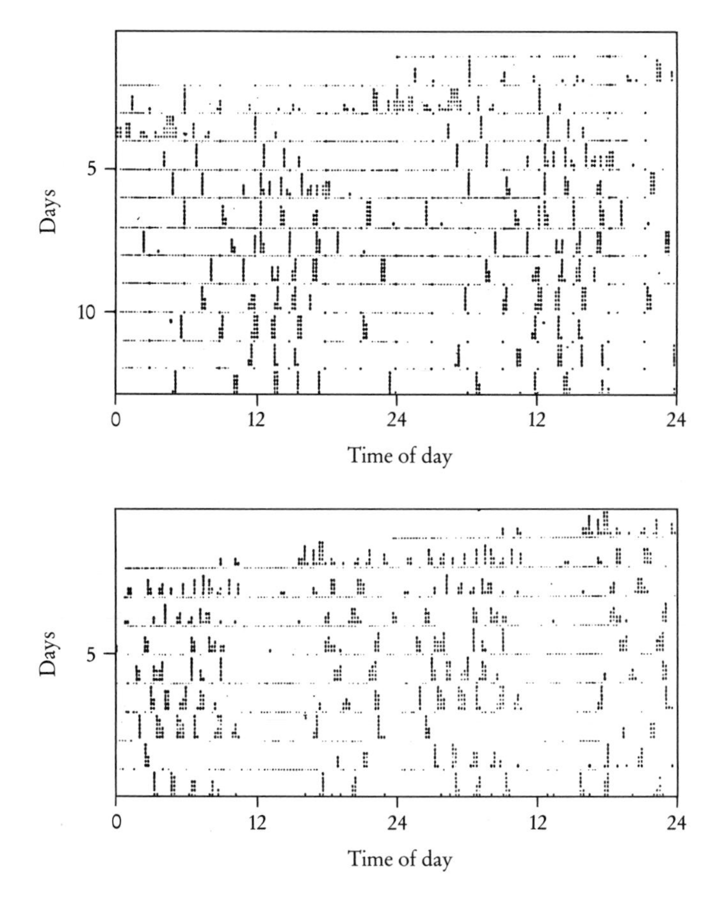

To reach the mosquito's singularity, about an hour of indirect daylight is required, applied at a moment that would coincide with midnight in the mosquito's summertime outdoor life. Kept in solitary confinement free of time cues after that peculiar experience, some mosquitoes—about one in five—behave very oddly. For weeks after, they are active in bursts every few hours, as though waking from catnaps. The regular alternation of rest and activity is replaced by perpetual insomnia. As in *Drosophila,* a second light pulse at any time cues the mosquito's clock to resume a predictable schedule.

Other individuals recover from the singular exposure spontaneously after a few days, as though from jet lag, but to widely diverse schedules. Still others—perhaps half—were normally rhythmic after only a day or so, also with an unpredictable phase shift. Does this show that *Culex* are not interchangeable clones, but individuals who exhibit both of the "limiting case" behaviors imagined as alternatives in the fruit fly, and intermediates besides? Maybe. Or maybe it only shows that individuals had slightly different clock phase or a slightly different orientation toward the light, so that some received a more nearly singular stimulus than others. It is a puzzling fact that the progeny of a single pair of a highly inbred laboratory strain, reared in the same bowl, monitored simultaneously at the same age under identical conditions, still show widely different daily activity patterns; even their periods range from 23 hours to 25 hours. No one yet knows whether these differences alone account for the different individual styles of singular arrhythmia, or whether individuals also differ in their rates of recovery from identically received stimuli.

Metaphorically, you can bump the clock in an individual organism to near equilibrium, just as you can with a pendulum or a child in a swing. A push on the swing when it is in motion will usually reset the phase of its rhythm. But a push of just the right size just in the middle of the cycle will stop it. Is the equilibrium stable? That depends on the child. If he's hyperkinetic, he'll pump it up again almost before you notice the pause. If he's more like a simple biophysical oscillator, his residual amplitude may grow exponentially at first, like money at compound interest. How fast it gets to the standard amplitude will depend on the "interest rate" of amplitude and also, delicately, on the initial principal, that is, on how nearly the original shove left the swing exactly at rest. In any case the phase of the new rhythm will bear no relation to the prior rhythm's phase. Biological clocks follow the same topologically imposed rules of phase resetting, but seem otherwise as diverse as children.

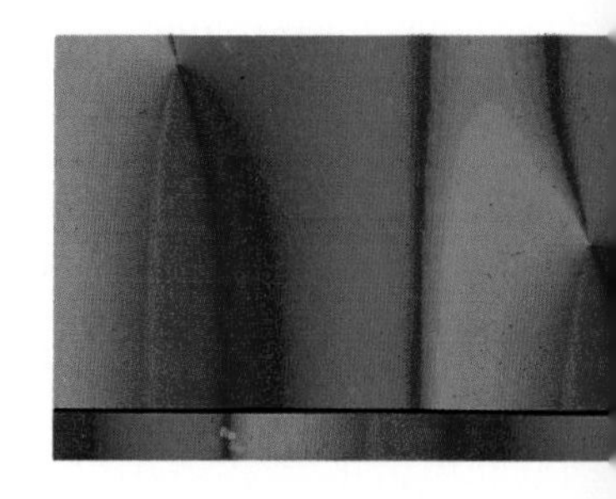

Someday, as part of a study of jet-lag insomnia, humans will systematically phase-shift themselves during temporal isolation. This experiment was in fact begun by C. Czeisler and E. Weitzman in New York in 1977,[5] but before the first light pulse was given, the subject astounded observers with the unprecedentedly long intervals of sleep and waking that we saw in Chapter 2, and the direction of research was deflected. The experiment is still necessary. Will our own species turn out to have a singularity too? According to the majority of mathematical models entertained today, we must have. The practical question is only what stimulus will evoke it. And how will an individual feel after that critical exposure to such oddly timed cueing? Will he be caught up in perpetual jet-lag drowsiness persisting for weeks? Or will he go about his work with abnormal vigor, no longer a slave to the relentless pull of the bed? The answers might be in before this book is published.

Cellular Clock Shops

Though the excitement seems naive now, there was a night in 1967 when I could scarcely sleep, before returning to my perpetually lightless "laboratory" in the storage closet of a basement animal room (this was before the discovery of safelights). Would metamorphosis go awry in the first singularity experiment? When the trays of hourly eclosion counts came out into the light and eyes readjusted to the glare, it was disappointingly clear that the flies had not failed to hatch, and they seemed in good health. Many biologists had felt that circadian timing might play a vital organizing role in the intricately timed reassembly of larval tissues into adult fly tissues; but here the circadian rhythm had been annihilated harmlessly.

Or had it been? True, the individual fly no longer harbored a normally functioning clock. But even if each fly (pupa) seemed devoid of rhythmicity, we might still pursue the question of incoherence one layer deeper. Circadian timekeeping clearly was not being suppressed by a nonpermissive environment: the darkness and temperature remained the same as for flies in adjacent populations that were emerging normally in daily bursts. No, this was a new mode of normal function in a healthy clock under the same conditions that alternatively permit regular circadian oscillation. From the fact that the fly or mosquito prefers no hour over another after seeing a singular stimulus, we learn only that the collective biochemical or electrophysiological rhythms of cells in its timekeeping tissue amount to nothing very rhythmical. Yet each individual might still be itself a clock shop, a playground of many swings, each still cheerfully swinging, just no longer in concert. How could we rule out this interpretation of the singular state?

Incoherence in a group of cellular clocks would not be so very strange. Most of the cells in your own tissues are operating incoherently as you read, at least in their cycles of cell division. In contrast to such permanent cells as the neurons of your brain, many other types of cells suffer a substantial mortality and must be replaced, usually from a population of stem cells. Epithelial cells such as those found on the surface of the tongue or the cornea of the eye are good examples. They divide predominantly in the morning, but not exclusively so; mitotic events do occur all around the clock. As noted in Chapter 2, this tendency to synchrony is a basis for substantial improvements in chemotherapy by deliberate timing of exposure to drugs. If cell populations were more synchronous, the improvements could be spectacular: but their circadian synchrony is only partial. What if circadian rhythms are *generally* somewhat incoherent at the cellular level? If a phase-resetting stimulus catches different cells at different old phases, it resets them diversely. Encounter with a near-singular stimulus sends clocks to quite different new phases even if their old phases were nearly equal. Such a perfectly placed cue-ball shot would scatter even a nearly synchronous collective of cells; and conditions for scatter are more easily met in populations that were already somewhat incoherent. It is therefore reasonable to ask how much of an animal's behavioral arrhythmia is due to extinction of every cell's private rhythm, and how much merely reflects loss of synchrony within the multicellular clock itself. We will not know until the circadian rhythms of many cells can be monitored individually in an organism made timeless by the singular stimulus.

Since the middle 1970s we have known what tissues to look at: in insects, a certain part of the brain; in some kinds of birds, the pineal gland (which is also a part of the brain); and in mammals (including primates, and probably our own species), a bilaterally symmetric cluster of nerve cells called the suprachiasmatic nucleus ("nut above the crossing") in the hypothalamus, just above the point where the two optic nerves cross en route to opposite sides of the brain (see "Brain Clocks," in Chapter 3). It may not be long before this question about unforced circadian arrhythmia, and so about the ultimate nature of circadian timing, can be asked and answered by direct observation of cellular rhythms. But for the present, an answer comes indirectly from the same test that was used at the grosser level of a population of pupae.

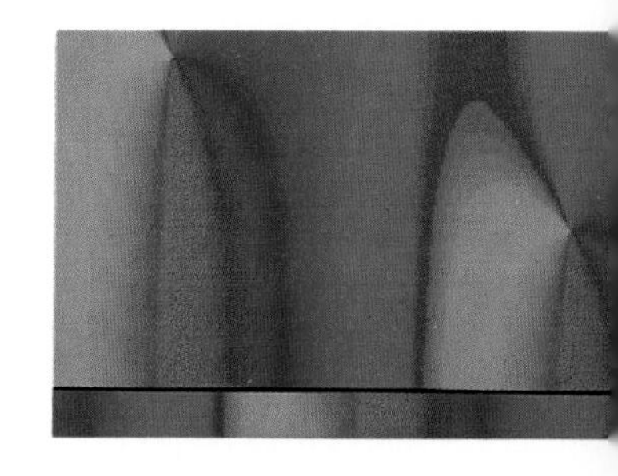

The experimental results in *Drosophila* and *Culex* are compatible with either interpretation: the individual may have a single, solidly coherent clock whose amplitude of rhythmicity can be reset, and is reset near zero by the singular stimulus—or its clock may be a collective of many little independent clocks whose timings are so scattered by that singular cue-ball shot that the aggregate rhythm is washed out. Probably both factors contribute, but their relative proportions have not yet been resolved experimentally. Another organism may give a cleaner answer.

The Sleep Movements of Plants

The question of cellular incoherence was first posed to a plant, the little red flower of *Kalanchoë blossfeldiana*. Historically, plants were the first living things to reveal an innate circadian rhythm. This was the daily opening and closing of mimosa leaves, assumed since the time of Alexander of Macedon (and perhaps by most people today as well) to be a response to the alternation of sunlight and darkness. That is correct, as long as there *is* an alternation of sunlight and darkness. But in the early 1700s, the French astronomer Jean Jacques de Mairan inquired further and discovered the plant's innate rhythmicity by the simple expedient of keeping the plant enclosed in a dark place: its leaves still opened and closed on nearly the same old schedule. A few decades later, Henri-Louis Duhamel du Monceau repeated this observation with greater care, keeping the plants in a wine cave at constant temperature as well. A century after the first observation, Augustin Pyramus de Candolle, in Geneva, repeated those experiments and additionally discovered that the innate rhythm can differ distinctly from 24 hours: during exposure to ceaseless dim light, the plants cycle regularly every 22 to 22.5 hours, thus excluding the plausible objection that some covert 24-hour influence was regularly cueing them. After another half century, Charles Darwin wrote a whole book, *On the Power of Movement in Plants,* vexed by doubt whether internal clocks could arise through his proposed evolution through natural selection. That was only a century ago, and Darwin's question remains open, as the selective advantage of *innate* rhythmicity remains obscure in most cases.

The familiar red flower of *Kalanchoë blossfeldiana* opens and closes in a circadian cycle even when plucked into a tiny vase of sugar water in temporal isolation.

As in many other plants, the *Kalanchoë* flower opens and closes every day. It persists even when plucked and placed in a tiny vial of sugar water. It reverts to its native period of 23 hours when placed in a darkroom—actually a "green room," comparable to the "red room" or "yellow room" required for *Drosophila.* Plants are blind to green light but the experimenter is not, so he need not fumble in utter darkness.

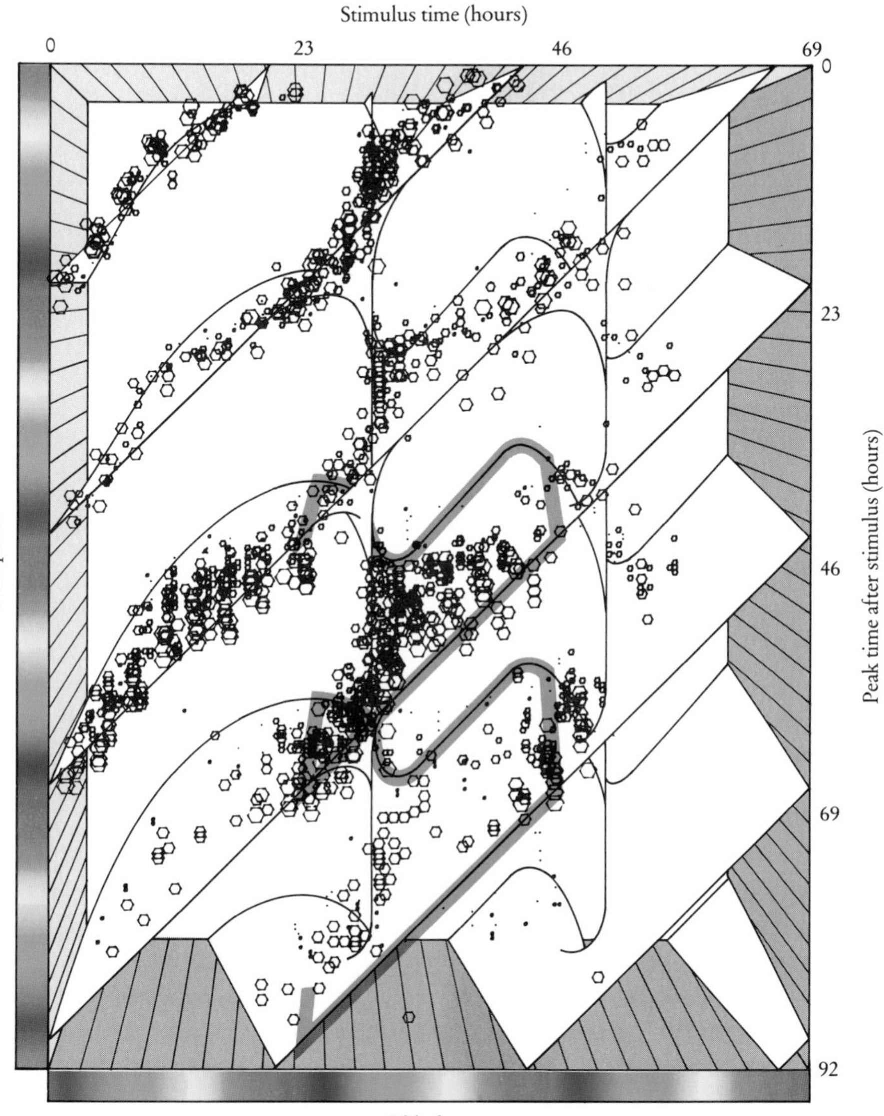

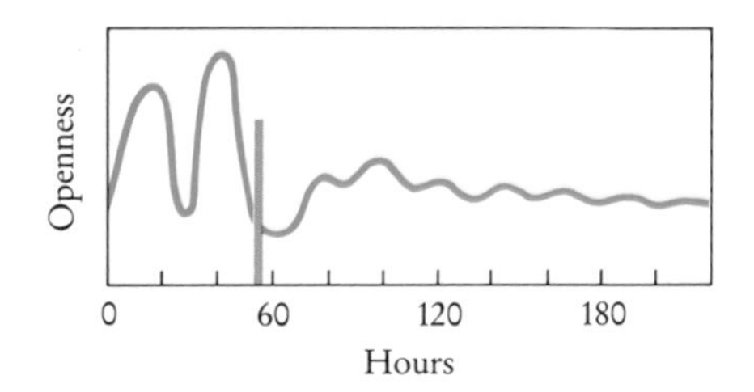

Left: The time crystal of the *Kalanchoë* flower clock, exposed to pulses of red light. The size of the stimulus ranges from zero (in the front) to three hours. A colored helix outlines two unit cells. The rhythm used for assay of old phase and new phase is the time course of the area shadowed on a photocell by the flower as it opens and closes. *Above:* Annihilation of the *Kalanchoë* flower rhythm by a singular stimulus, 120 seconds of red light at subjective midnight.

When he exposes the little flower to bright red light, however, its clock resets in the pattern found in organisms with clocks in their brains. In Tübingen, Germany, Wolfgang Engelmann and collaborators Anders Johnsson and H. G. Karlsson, visiting from Sweden, repeated the protocol used for *Drosophila:* they measured *Kalanchoë*'s time crystal.[6]

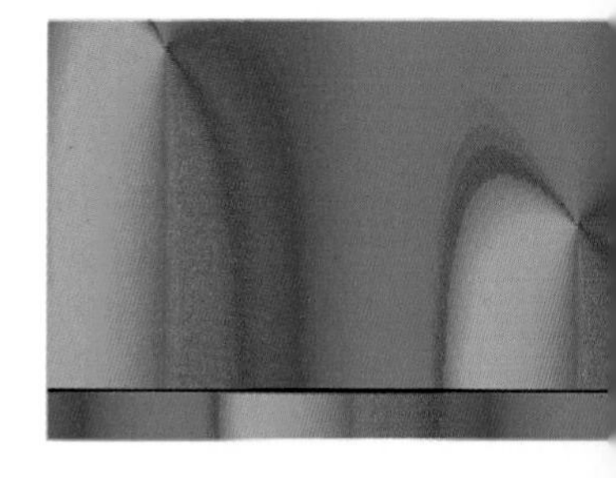

Like the fly, the flower goes arrhythmic after a single exposure of the right duration at just the right moment, and that moment recurs at subjective midnight, when the flower is most fully closed, every 23 hours. Thereafter, it no longer opens and closes. As in fruit fly pupae, there is still no indication of spontaneous recovery a week later.

How is the arrhythmia of *Kalanchoë* manifest on the cellular level? This time, an experiment suggests an answer. Suppose (A) the arrhythmic flower is composed of normally rhythmic single cells, randomly reset. This is certainly possible. Experiments have shown repeatedly in diverse plants that circadian rhythms are innate in many parts of the plant, probably even in single cells. The plant, then, less resembles a solitary clock than a clock shop in which all clocks are normally set by the light/dark cycle to which all are normally exposed. Suppose, then, that after the singular stimulus each flower is an asynchronous clock shop. A light pulse too delicate to do much phase resetting will not then resynchronize the separate clocks enough to muster uniform circadian control over the timing of petal movements.

In contrast, suppose (B) the cells are not oscillating at all, but rather are all delicately poised on some kind of equilibrium. Then a pretty slight nudge may be sufficient to kick them off it. The rhythm so initiated, if normal, will have a predictable phase. Engelmann and friends tried it. The result: a sufficiently brief red exposure (about one-fourth of the singular duration) only slightly resets the cellular clocks of a normally rhythmic flower, and so it should not be able to substantially resynchronize a clock shop of diversely phased normally rhythmic cells; but it restores rhythmicity to flowers previously made arrhythmic by the singular stimulus. This contradicts (A) and agrees with (B). The arrhythmia of the single flower appears to be arrhythmia at the cellular level as well. Corroborating this indirect inference, Engelmann checked the diameters of individual petal cells. Normally they vary as turgor pressure (driving the petal movement) waxes and wanes. There were no significant variations after the singular pulse, again incompatible with the idea that cellular clocks have only been reset randomly.

This conclusion requires further corroboration, but deserves a place of honor here as the first and to date perhaps the only deliberate effort to give an experimental answer to these pivotal questions in clock physiology: Do the cells of a tissue typically keep time independently? Can the circadian mechanism be shut down nondestructively in single cells by a unique fleeting tap that would do nothing unusual if delivered more delicately or more vigorously or at any other moment?

A Unicellular Clock

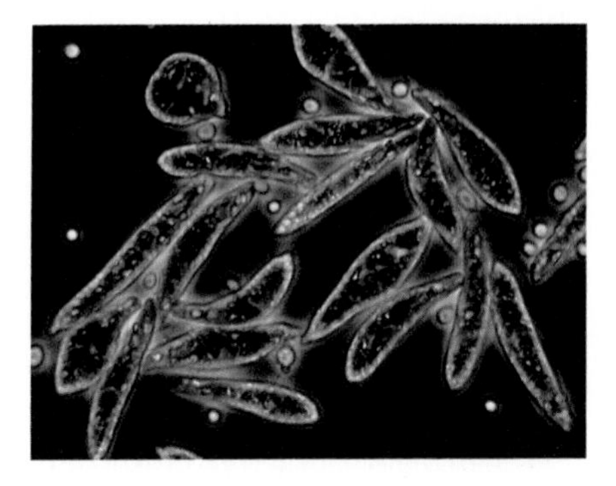

The clock's phase singularity has been sought in only two unicellular organisms, *Gonyaulax* and *Euglena*: it has been found in each. *Euglena* is motile like an animal but is green with chlorophyll for photosynthesis. Its reproduction by binary fission is gated by the internal circadian clock.

The second question may be resolved first in another plant, the free-swimming unicellular alga *Euglena.* The rate of random swimming in *Euglena* is greatest in the daytime and least at night, when cells tend to settle to the bottom of a column of water. Cell division is reserved for night also: cells mature over the course of several days toward readiness to divide, but when each finally does, it divides during a window of hours repeatedly opened by the circadian clock. Even when light is available during both day and night, *Euglena* still divides only during intervals spaced 24 hours apart, and it persists in its circadian rhythm of motility as well. In the dark, as with most other circadian systems, *Euglena*'s internal clock can be reset by exposure to a single pulse of light. If the exposure is bright, then new phase depends on old phase in the even pattern, whereas a negligible exposure elicits odd resetting (necessarily, since new phase = old phase). Moreover, a single pulse of some critical intermediate intensity proves to abolish the rhythm, if and only if applied in a vulnerable phase near the end of the usual interval of cell divisions. Thereafter the population continues to multiply, but in a smooth exponential way without the familiar circadian alternation of permitted and forbidden intervals.[7]

Is this because every cell has turned off its circadian clock, or only because each responded idiosyncratically to that unique stimulus, and resumed its rhythm with random phasing? It seems possible to find out by looking at the vertical distribution of cells in a column of water. If the circadian clock persists, randomly phased, then the top strata should be enriched in cells whose clocks are currently at the vigorous-swimming phase of their cycle. By siphoning off the top strata, a culture would be isolated that exhibits rhythms of motility and of cell division, predictably phased. In contrast, if all clocks were quiescent then cells at any depth would be the same and utterly arrhythmic. Answers may be in before this book is published.

Another likely candidate for resolving this question is the unicellular plant *Gonyaulax.* Its biochemical activities—photosynthesis, cell division, bioluminescence—are segregated according to clock time, as though the cell were punching in for different jobs on successive shifts.[8] Normally, these activities are synchronous throughout the population, not because cells communicate to keep in step, but only because each cellular clock is independently set every morning when sunlight streams down into the cool blue waters. These rhythms can be monitored in the

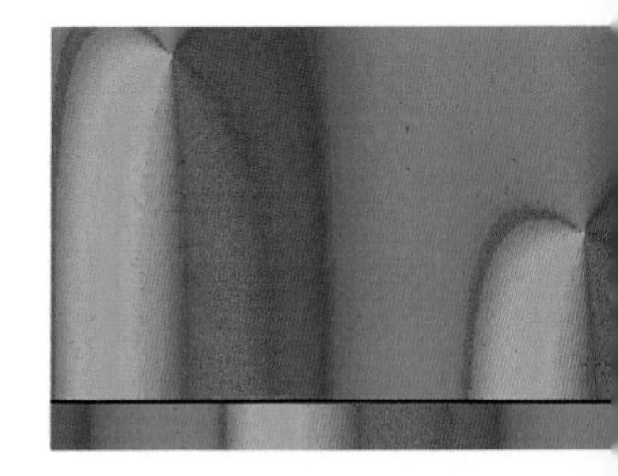

individual cell, but since all the *Gonyaulax* cells in a vial of seawater are synchronous anyway, it is usual to experiment with vials of about 50,000 cells.

Frank Sulzman and colleagues in the Harvard Biological Laboratories first enquired whether *Gonyaulax* cells synchronize one another when removed from the synchronizing sunlight.[9] They mixed cell suspensions from different time zones. Each experiment began with separate vials of seawater in which the little blue-glowing clocks were set to Boston time, Denver time, Hawaii time, and Tokyo time (or more exactly, 2, 6, and 12 hours from Boston time). Then the vials were taken from their light/dark cycles to the constant conditions of the glow monitor. Part of Hawaii was poured into part of Tokyo, for example, and all three glow rhythms were carefully recorded. Various proportions of different suspensions were mixed in various experiments. What compromise did they reach? None at all, surprisingly: the two intermingled populations persevered in their innate rhythms as though still kept in separate Hawaii and Tokyo vials.

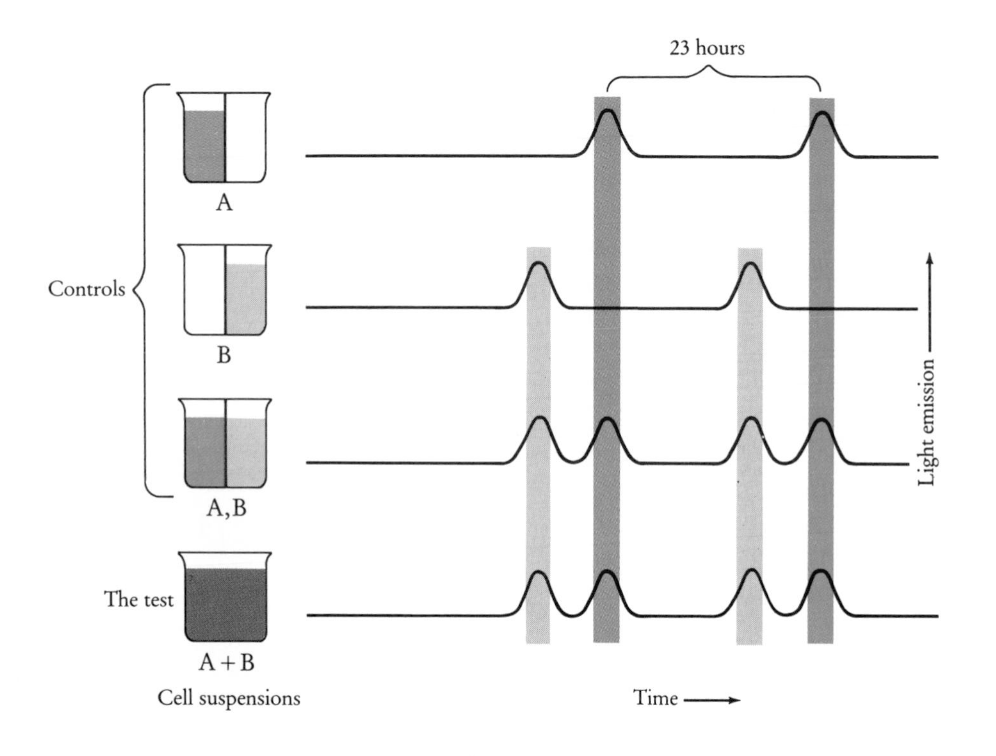

Cultures of *Gonyaulax* are set to different time zones (A) and (B). In (A, B), light from separate cultures is recorded simultaneously: it is indistinguishable from the sum of (A) and (B). In (A + B), the cells are allowed to intermingle to see whether they still glow independently or reach a compromise timing. In the first four days of mutual exposure, the result is not reliably distinguishable from control (A, B). The cells are independent, to first approximation. More prolonged mutual exposure indicates some slow mutually synchronizing interaction.

This 1982 result was surprising enough to deserve another try three years later by J. "Woody" Hastings and others, in his Harvard laboratory.[10] This time, cultures were mixed in roughly 2:1 ratio with nearly opposite phases and followed, not just for three or four days after mixing, but for another week longer. The results were essentially the same for the first few days, but during the second week of this marathon the two glow peaks seen in the mixed culture gradually fused to one: the minority culture adopted the timing of the majority. However, if the medium was changed every two days, the two cultures failed to synchronize. Their interaction was evidently mediated chemically through the seawater.

What is happening in a soup of *Gonyaulax* comfortably situated in a Boston laboratory darkroom where the sun never rises? If cells are not entrained by diurnal rhythms from outside, they do not vigorously entrain one another's clocks, but do they communicate well enough to stay partly synchronous? Constantly fed a little light, just enough to keep alive, the cells emit blue light in a spontaneous rhythm containing distinct peaks. Go away and come back to look a week later: the glow peak still recurs every day, but about 7 hours off the schedule of the surrounding lab and city. The rhythm of glow does not repeat at 24-hour intervals, but at intervals of 22 hours and 52 ± 2 minutes, dependably losing 68 minutes a day relative to outdoor time.[11] So there has been no slip-up: the cells are not following some covert influence (a leakage of light into the laboratory, traffic vibrations, who knows what?) from the 24-hour throb of Cambridge outside the darkroom. They are indeed marching to their own drummer, and quite precisely. But as time passes there seems to be a little loss of synchrony between individuals.

How little? The "± 2 minutes" refers to the center of the peak of the daily bioluminescence. Remember, we are looking at the collective glow of tens of thousands of independent cells. After scores of cycles they could be expected to drift substantially out of step, broadening the daily glow peak. In fact, after only a week, the daily glow peak is detectably broader. David Njus, as a graduate student at Harvard, measured the broadening and found that cells differ in period by less than 4 percent (20 minutes to an hour out of 23 hours, depending on whether you mean "differ consistently" or "differ irregularly").[12] Possibly by mutually synchronizing one another to some degree, cells in a population achieve greater precision than they would in isolation, but it is clear that they do gradually disperse anyway.

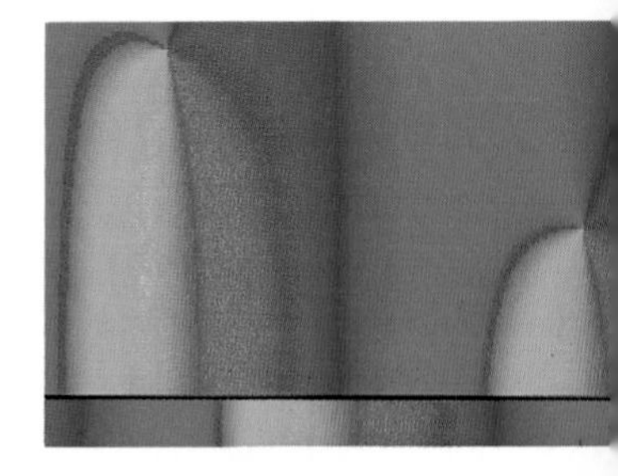

Why they are so precise remains a mystery. The mean period in *Gonyaulax* is about an hour shorter than the earth's, *Drosophila*'s is right on, and our own is an hour longer. Given an entraining environment, the exact native period seems relatively unimportant. In nature, daily sunlight resets these cells without fail to the "master clock," the earth's rotation. In the course of millennia, mutants must have arisen whose timing is off by an additional half an hour or so one way or the other each morning. Where are these mutants? Are they hidden by chemical entrainment to the rhythm of surrounding normal cells? Whatever the reasons and mechanisms may be, the tight synchrony of *Gonyaulax* individuals makes them convenient to biochemists eager to measure phase resetting by chemical intervention.

The Time Crystal of *Gonyaulax*

Because *Gonyaulax* is a single cell, its physiology is still not practically distinguishable from biochemistry. Thanks to the pioneering efforts of Woody Hastings at Harvard and Beatrice Sweeney at the University of California at Santa Barbara, *Gonyaulax* became the first favorite organism for biochemical inquiries into the clock mechanism. These inquiries have largely taken the form of clock perturbations: the cell is fleetingly exposed to some metabolically significant substance to see whether that molecule interferes with the clock mechanism. The most reliable assay of brief interference is phase resetting. *Gonyaulax* reacts by even resetting to various metabolic inhibitors, of which the most studied are protein-synthesis inhibitors because that part of normal biochemistry seems implicated in the clock mechanism. Brigitte Walz and "Beazy" Sweeney[13] at Santa Barbara found even resetting in *Gonyaulax* and speculated about the implicit singularity. Walter Taylor and collaborators[14] in the Biological Laboratories at Harvard used the protein-synthesis inhibitor anisomycin to search for it, following the format of the original fruit fly experiment.

They found the singularity with a critical dose of the drug, at a vulnerable phase near the cells' subjective midnight, as usual. The resetting time-zone map is shown in color on the facing page. Phase measurements were difficult around the singularity because the rhythm was unpredictable and often obscure. In fact, they occasionally obtained flatter traces than ever seen before from any circadian clock, following a pulse in this region of ambiguity.

Does this result indicate a scattering of cells to all time zones, or does it indicate a turnoff of every cell? The question has not yet been an-

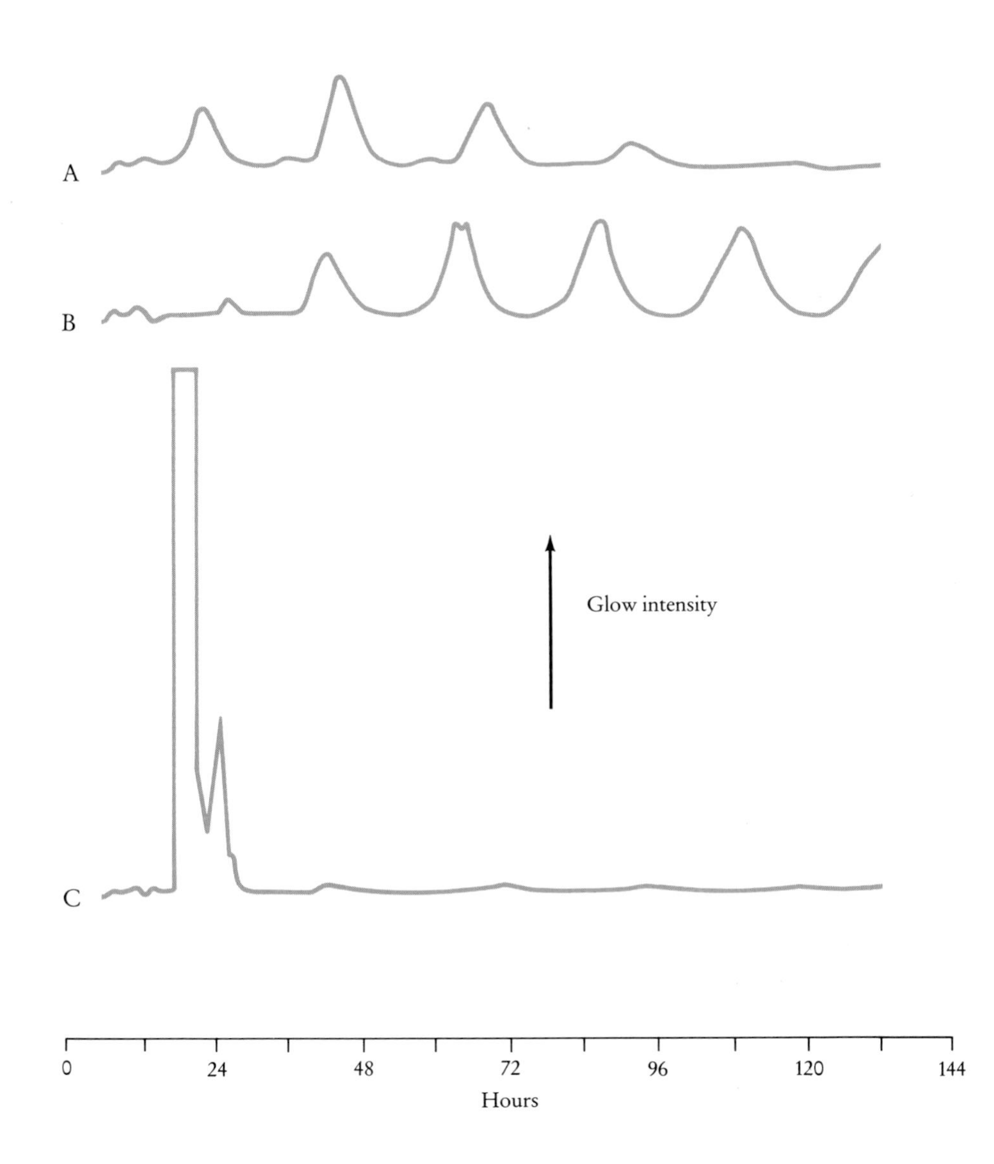

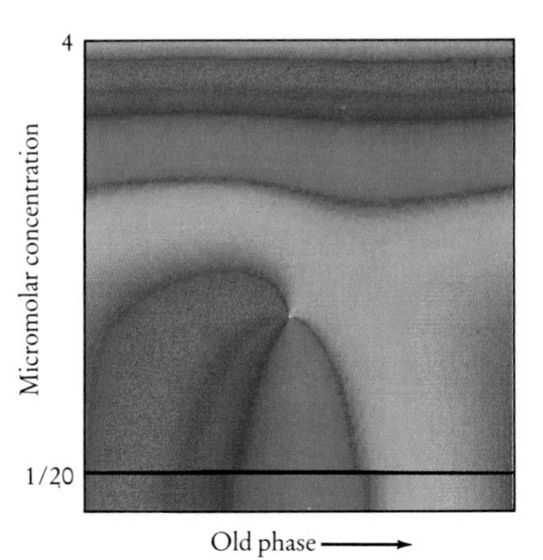

Left: The glow rhythm of *Gonyaulax:* (A) an unperturbed control; (B) the typical rhythm following any stimulus other than the singularity is merely phase reset; (C) after the singular stimulus, the glow goes off scale then stays flat at a level below the full-cycle average of traces (A) and (B). *Above:* Phase-resetting surface for *Gonyaulax.* The stimuli were one-hour pulses of anisomycin in various concentrations, applied at various old-phase moments in the circadian cycle. New phase is color coded. The concentration scale is logarithmic, starting from 1/20 molar, not 0, so the bottom of the plot is not quite identical to the phase ruler.[15]

swered at the single-cell level, but the present circumstantial evidence is strong. In the few vials that became strikingly arrhythmic, an unexplained burst of light followed the singular annihilating pulse, then the glow trace stayed day after day near the background level that is usual between daily peaks of glow. The collective peaks have not just been smeared out by scattering cells to all time zones; they have vanished. The cellular clocks were not reset: they were bumped to equilibrium like the cuckoo clock.

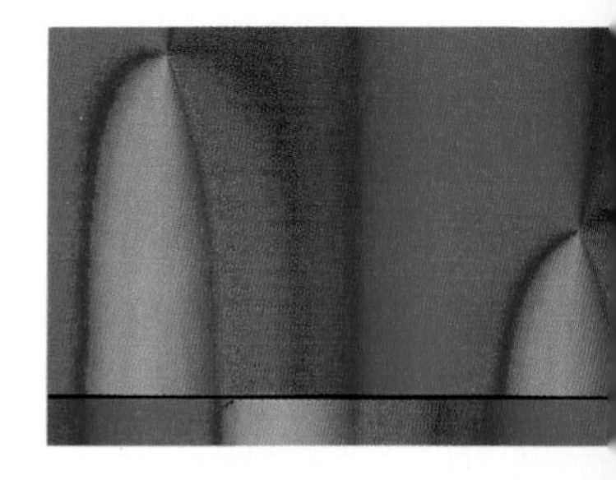

Amplitude

This chapter has addressed the decade-old apprehension that phase singularities might after all reflect nothing about the mechanism of circadian rhythmicity, but be only a consequence of summing diversely phased rhythms over populations of cells. This scruple could still prove meaningful in the neuro-humoral clocks of higher vertebrates, but the results of experiments with insects and plants appear to dismiss it. The implication is that singular arrhythmicity is due to an aberration of the strength, vigor, or amplitude of the circadian clock. The amplitude of a clock is apparently as resettable as the phase. How long it stays reset is now the question at stake, and the alternative answers can now be reformulated in these more sophisticated terms:

Suppose that clock amplitude remains depressed for a long time after a near-singular stimulus. Then we may attribute singular arrhythmia to the newly discovered fact that the circadian clock has amplitude as well as phase, and amplitude, like phase, is stably reset by a stimulus, even to zero: each cellular clock is turned off. In terms of children in swings, all swings were exactly synchronous, all were bumped in the same way, completely eliminating forward momentum at the bottom of the arc, and all remain inert thereafter until someone gives a push.

Suppose, on the other hand, that clock amplitude is stable only along some standard cycle, and rebounds immediately from a depressed situation near the singularity. Then we may attribute the observed persistent arrhythmia following a singular stimulus to a scattering of phase among separate clocks, each recovering from a transient unstable condition of zero amplitude and correspondingly ambiguous phase. Instances of immediate recovery to normal rhythmicity (at unpredictable phase) may indicate that there is only one clock, or that physically separate clocks interact too strongly to recover at widely dispersed phases. In terms of children in swings, each child is hyperkinetic, and pumps back up without delay, possibly while watching her companions to stay in synchrony. The timing of the recovered oscillation depends delicately on residual vibration when pumping began.

Intermediate cases between these extremes are probably common, in which arrhythmia soon after the singular stimulus represents uniformly low amplitude in all clocks, and later arrhythmia, or later development of a collective rhythm of unpredictable phase, is a consequence of their gradual recovery to normal amplitude, either independently or to some extent cooperatively. In terms of children, they are slow to pump up again, and possibly slow to get into mutual synchrony.

One consequence of adjustable amplitude was considered in Chapter 2. The character of control over the timing of inescapably discrete events (periscope breakthrough, cell division, eclosion) switches at a critical amplitude from smooth modulation of timing to discrete packaging of events into a daily "window."

Another consequence of adjustable amplitude is that a resetting stimulus changes the shape of the subsequent resetting curve itself. Reaction to a stimulus must depend on clock amplitude. To take the most extreme example, at zero amplitude the reaction to a stimulus must be the same regardless of timing, so the resetting curve should be (and is) absolutely flat. In general, the result of a second stimulus depends not just on the prior phase shift due to the first one, but also on the reset amplitude. The persistence of this effect has been demonstrated experimentally in great detail in the fruit fly,[16] but has yet to be tested in other organisms. To do so is important for understanding entrainment, the original reason for interest in phase resetting. The most successful analyses of entrainment all suppose that after a stimulus ends, any effects other than phase resetting fade away before the next stimulus arrives. That may be so following artificial laboratory stimuli that are brief enough, small enough, or large enough. But it seems not so in the middle ground.

It will be essential to find out where that middle ground is for the natural stimuli of importance to any particular circadian rhythm. For example, the day may come when travelers go to the medicine cabinet before a long east-west flight and select a capsule marked "San Francisco to Beijing: take one at California noon before leaving and one at China noon after arrival." Workers in a permanent space station will also want to adopt the phase of a target city before riding the shuttle down for a vacation at earth gravity; and during the last days of vacation they will need to resynchronize to their work schedule in the station. This will probably entail a short sequence of resetting stimuli, perhaps provided pharmaceutically in those capsules, or by measured exposure to bright light. To prescribe the optimum dosage will require knowledge of the clock's phase *and amplitude* at the moment of each dose.

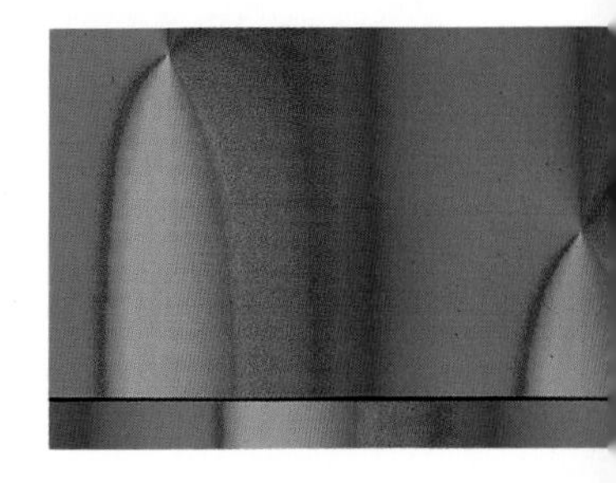

The yellow slime mold *Physarum polycephalum* contains nuclei that divide rhythmically and all in synchrony as the organism grows and spreads. In these nuclei, the chemical mechanisms of phase control implicitly involve discontinuities or singularities of timing.

Chapter Seven

The Singularities of Biochemical Clocks

About thirty years ago there was much talk that geologists ought only to observe and not theorize, and I well remember someone saying that at that rate a man might as well go into a gravel pit and count the pebbles and describe their colors. How odd it is that anyone should not see that observation must be for or against some view to be of any service.

CHARLES DARWIN (from *THE ORIGIN*, by IRVING STONE)

Swarms of blue-flashing *Gonyaulax* are not the only known populations of self-sufficient and surprisingly precise circadian clocks. Populations of fruit flies maintain comparable coherence: pupae differ only about 1 percent in their 10-day-average periods.[1] Diverse bits of tissue from various plants, fungi, insects, crustacea, snails, and rodents maintain their circadian rhythms in laboratory culture vessels, isolated from all known timing cues.[2] No one yet knows to what degree cells in clock tissues affect one another's circadian clocks, and with what consequences for collective temporal organization. In temporal isolation, the parts of the plant still exhibit circadian rhythms of respiration, of leaf position, and of flower opening. After many cycles in constant conditions, however, parts begin to get out of step with one another. The petal movements of composite flowers, for example, get progressively more and more desynchronized. A flower that initially opened and closed its parts in concert eventually gives a grotesquely uncoordinated performance, not because parts have ceased to open and close with a 24-hour rhythm, but because they do so out of step.[3]

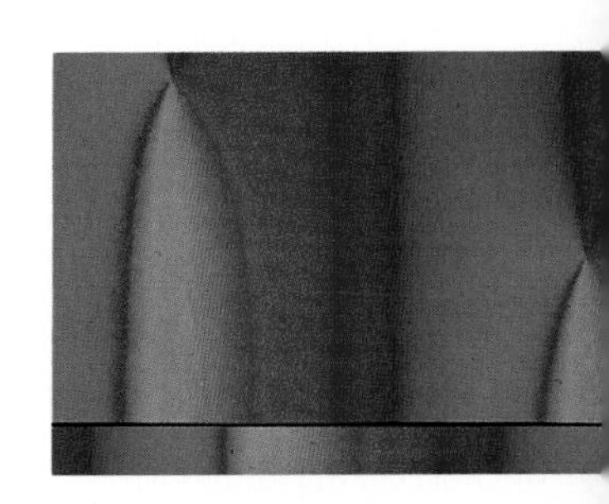

It seems that the cells of green plants lack the means to entrain one another, perhaps because mutual synchrony has always been guaranteed anyway by daily exposure to the solar cadence caller. But the cells of a fungus or a rat do not directly depend on the sunlight for their sustenance. They can survive for months, even years, in the lab without exposure to daylight. What keeps them synchronized? That at least some cells do stay synchronized is revealed, for example, in the persistence of the monkey's sleep/wake cycle after years in Richter's laboratory, unaffected by the rhythm that drives the rest of the world. Evidently the animal's clocks—the bilaterally symmetric halves of the suprachiasmatic nucleus and maybe the individual cells within each—keep one another mutually synchronized. The suprachiasmatic nucleus controls rhythmic secretion of the brain hormone melatonin by the pineal. Rhythmic injections of melatonin can entrain circadian clocks. By controlling melatonin secretion, then, the circadian clocks in brain cells may entrain one another. Rhythmical secretion of a chemical cue to which clocks are rhythmically sensitive for phase resetting is a common method of mutual entrainment. Recent experiments with *Gonyaulax* have hinted at it, though many cycles passed before the effect was even detectable.

At the opposite extreme of mutual coupling strength are the biochemical clocks of yeast cells, which keep one another in strict lock-step. But even in populations that thus vigorously discipline any wayward cell-clock, ability to maintain synchrony in the face of a misplaced time cue has limits. One particular limit is impossible to evade by biochemical or biophysical refinements, being independent of mechanisms: this is the topologically necessary existence of a phase singularity.

The inevitability of phase singularities, even among mutually synchronizing cells, is particularly transparent in the timing of energy transfers in yeast cells. In this chapter we deal, not with circadian rhythms, whose biochemical mechanisms are only now being deciphered, but with the best-understood biochemical clock.

Rhythmic Energy Metabolism

The want of this incomparable artifice [the microscope] made the ancients . . . err in their . . . observations of the smallest sort of creatures which have been perfunctorily described as the disregarded pieces and hustlement of The Creation . . . In these pretty engines are lodged all the perfections of the largest animals . . . and that which augments the miracle, all these in so narrow a room neither interfere nor impede one another in their operations. Ruder heads stand amazed at prodigious and collossean pieces of Nature, but in these narrow engines there is more curious mathematicks.

HENRY POWER, *EXPERIMENTAL PHILOSOPHY,* 1663

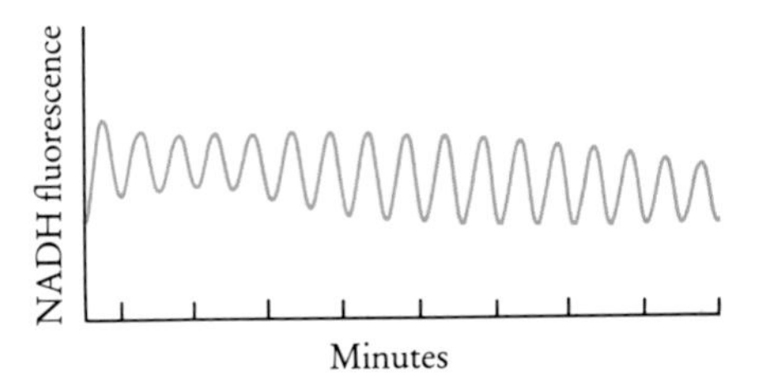

NADH fluorescence in a suspension of yeast cells shows 30-second oscillations after reserves of dissolved oxygen are depleted.

During a vital stage in the manufacture of beer or wine, it is essential to keep air out of the brew. Given oxygen, the yeast will chemically degrade sugar to vinegar, but without oxygen it uses a much less efficient biochemical pathway that stops short of acetic acid, at alcohol—an arrangement that mankind learned early to foster. Cells split sugar in order to extract a small profit from each transaction. The profit is energy, coined as a chemical bond—mostly, the third phosphate bond of adenosine triphosphate (ATP), but also as the energy-rich molecule nicotinamide adenine dinucleotide (NADH). NADH happens to fluoresce nicely in ultraviolet light, a circumstance first exploited in the late 1950s and early 1960s to directly follow the progress of energy metabolism by watching this blue-green glow. Observation: in the absence of oxygen, it waxes and wanes twice a minute as yeast cells transduce energy like alternating-current dynamos, at a pace 2000 times more leisurely than the engineered machinery of a 60-cycle regional power grid. Inference: just as in the power grid, these tiny engines keep each other synchronized, glowing and fading in unison under the ultraviolet lamp.

The Pasteur Effect

Why does anaerobic sugar metabolism not proceed in a steady way? An answer can be given by solving the known equations of regulation within this web of interacting enzyme reactions. But a loose and qualitative answer is also possible. Glycolysis is the main energy-producing degradative pathway in cells that lack oxygen or that lack mitochondria, the subcellular organelle of respiration and home of the Krebs cycle pathway. Such cells include your red blood cells, the transparent cells in the cornea of your eyes, cells in poorly vascularized tumors and embryos, the microorganisms that produce yogurt, sauerkraut, and gangrene, and those that spoil food. In cells that use oxygen, the Krebs cycle, with several reactions that follow it, is like an afterburner; it additionally derives 18 times more ATP by oxidizing the residual carbon compounds from glycolysis. Overall, glycolysis cycles NADH and

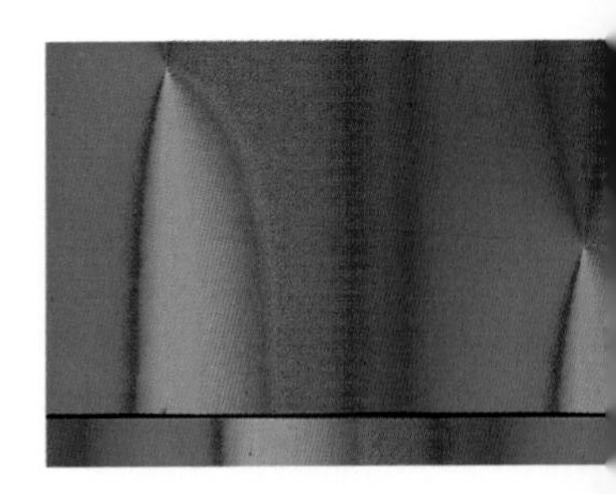

produces some ATP; the Krebs cycle and what follows produce much more ATP and *consume* NADH. Thus, in the presence of excess oxygen, ATP levels are high and NADH levels are low; there is no oscillation. If oxygen is absent or scarce, glycolysis predominates and ATP levels are lower, NADH higher.

In yeast and other cells that use oxygen, the rate at which sugar passes through the glycolytic pathway is controlled at the third enzyme in the pathway, the phospho-fructo-kinase (PFK) that attaches a second phosphate group to the fructose sugar molecule. When the cell's adenosine phosphate pool (consisting of mono-, di- and triphosphate derivatives) consists mostly of the relatively energy-poor adenosine mono- and diphosphates (as happens when mitochondria lack oxygen to work with), PFK activity increases as much as a hundredfold. In the absence of those energy-poor phosphates (when they have been upgraded to the triphosphate ATP by the Krebs cycle and so on), PFK shuts down. This shutdown by oxygen is called the Pasteur effect.

When PFK is active, it is powered by the triphosphate ATP and nibbles a phosphate group off, degrading it to a lower-energy species—which is to say, PFK manufactures its own activators! The more PFK activity, the more more. Here lies the potential instability of a steady flux. This instability develops into a bounded oscillation. The graphs on this page depict this cycle by plotting measured NADH concentrations against simultaneously measured concentrations of ATP and glucose-6-phosphate. Many other intermediates in the processing of sugar also reach their peaks and troughs one after the other while fluctuating above and below average levels.

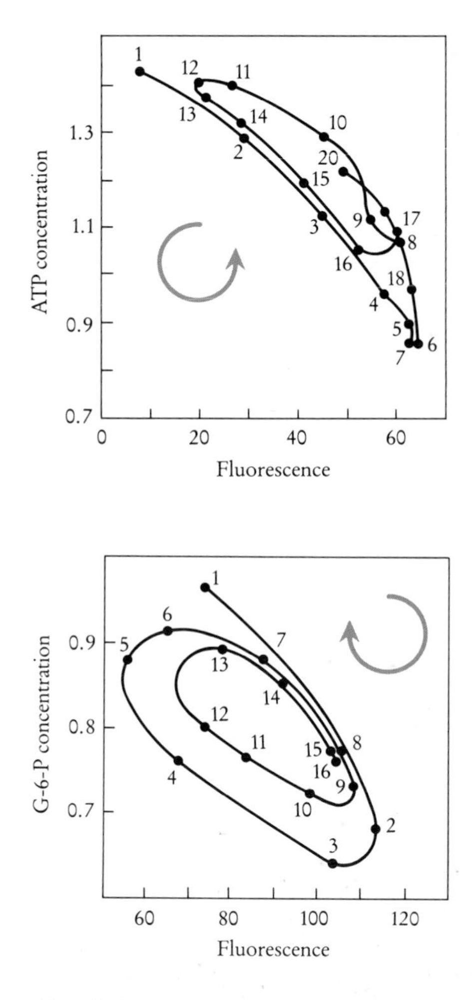

Chemical concentrations can be measured throughout the cycle by suddenly freezing samples taken from the oscillating suspension, then assaying them later. In these plots, each sample was measured for NADH concentration (monitored by the brightness of fluorescence) and for the simultaneous concentration of ATP or glucose-6-phosphate (G-6-P). From moment to moment in the cycles, all quantities vary rhythmically, but NADH reaches its peak (and its trough) about a half cycle before the others.

Time Crystals in Energy Metabolism

The yeast cell is a biochemical clock. Like other biochemical clocks (for example, the *Gonyaulax* cell), it responds to chemical interventions that engage its fundamental mechanism. Usually, the only permanent result of any such perturbation is a phase shift. E. Kendall Pye and I tested the effect of a tiny oxygen pulse, activating the cells' mitochondria to make some ATP and oxidize some NADH, to see how far the principles learned from *Drosophila* could be generalized: did they pertain only to circadian clocks, or were they about a large class of biological clocks?[4] This measurement was organized exactly like the pinwheel experiment: the rhythm was started by releasing a population of clocks from suppression by constant oxygen; the rhythm thus initiated was monitored in constant conditions; a stimulus was administered by momentary application of oxygen at each phase in the cycle; and the reset phase was recorded after stimuli of every pertinent size.

It might seem reasonable to expect even resetting, because a big enough dose of oxygen shuts down the oscillation altogether, turning down PFK activity and holding NADH at minimum levels, regardless of initial phase when the Pasteur effect is thus invoked. This is indeed exactly what was found. By repeating the singularity trap protocol, the implicit phase singularity was sought. It was found at 8 micrograms of oxygen per gram of wet cells, just after the NADH maximum. Below is the time crystal of the yeast cells, viewed from the big-stimulus side. You can see the even resetting curve in the foreground, the diagonal

The response of oscillating glycolysis to a discrete perturbation is plotted here three-dimensionally as a time crystal. The left-to-right axis measures the time of an oxygen pulse in seconds after prior NADH maxima. The data are repeated in a second cycle to the right. The vertical (downward) axis measures the time of subsequent maxima in seconds after the pulse. Pulse size is indicated in depth from zero in the background to a near-saturating dose in the foreground. Outlined in color is one turn of a helix where the resetting surface encounters edges of the unit cell.

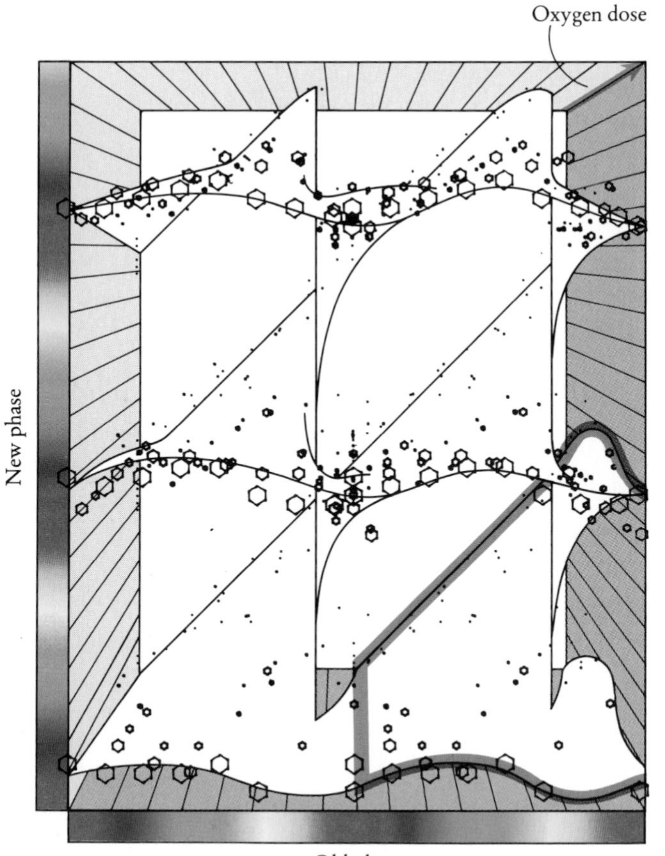

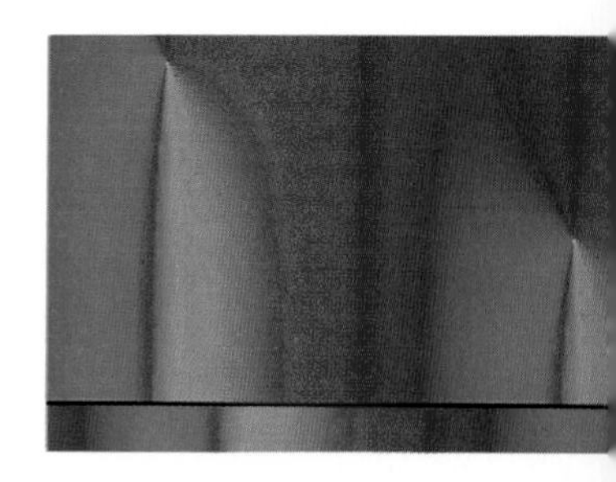

where new phase = old phase in the background, and in between, the screw surface with its singular axis. The axis points to a special stimulus on the plane of oxygen dose versus old phase. After this singular annihilating stimulus, the yeast suspension metabolizes glucose in an unstably steady way—the fluctuations are weak and irregular.

The screw-crystal pattern is thus not limited just to fly circadian clocks reset by a flash of light, nor to the circadian clocks of multicellular organisms reset by light, nor only to circadian clocks, whether in animals, plants, or single cells. The screw-surface-cum-singularity is not just a hallmark of the circadian clock mechanism: it seems to be common to biological pacemakers, even in the timing of heartbeat and breathing (page 97). In sugar metabolism, the screw-crystal pattern is also not limited to perturbations by oxygen: Leonard Greller, a Ph.D. student of E. Kendall Pye at the University of Pennsylvania Medical School in 1977, repeated the experiment by intruding upon sugar metabolism with acetaldehyde instead of oxygen: same results. Pye and his student John Aldridge upset the timing with calcium and magnesium ions: similar time crystal. Something deeply fundamental to the organization of biological timing is betrayed here by the persistent reappearance of the screw-crystal signature.

A Rotating Wave of Fluorescence

The contour map, on the facing page, of the yeast resetting surface is color coded by our usual scheme. As before, red codes the timing of rhythm maxima, extrapolated back to the moment of intervention. The colors thus code for the time zone to which the cells were reset by a pulse of oxygen. But the cells in each time zone continue to oscillate. Thus one time zone after the next will reach its maximum of fluorescence in order around the singularity like the rotating wave of hatching in the pinwheel experiment with *Drosophila* described in Chapter 5.

The rotating isochron in this many-part experiment is not really a wave in the usual sense, since nothing is really propagated or conducted from vial to vial. It is only an optical illusion due to the organization of timing experiments in space. Because the glow reappears at equal intervals everywhere, the wave "rotates" in the same period at any distance from the pivot; therefore it "moves" with no single characteristic speed. In comparable pinwheel experiments with circadian clocks, the wave would not be blocked by cutting a "firebreak" of dead fly pupae. But metabolizing yeast cells, unlike pupae, are capable of forcing one an-

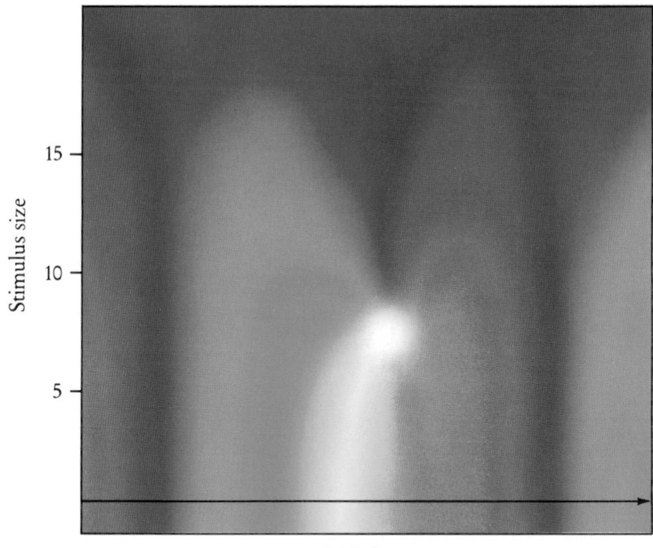

A color-coded contour map of the yeast resetting surface. A region of any single color contains stimuli (dose of oxygen and old phase when it is administered) that result in similar timing of subsequent NADH peaks, measured from the moment of oxygen injection. The critical dose (about 8 micrograms per gram of wet cells) following which timing is ambiguous must be given five seconds before NADH maxima (old phase "red").

other to synchrony. Thus a continuous suspension of yeast cells arranged in time zones might constitute a medium for the propagation of real chemical waves.

Phase Compromise Between Coupled Clocks

Phase resetting by oxygen is measurable only because we can work with a population of thousands of cells, and we could depend on them to glow and fade in unison cycle after cycle. Why do they stay together? Do two nearby cells arrive at a compromise to get in step if their clocks are initially set differently? The question can be posed as an experiment with two equal populations, initially in different time zones, suddenly mixed to surprise one another. In such an effort to detect communication between *Gonyaulax* cells, no compromise was soon struck; the cells almost wholly ignored one another. Yeast cells, in contrast, behave very differently; they synchronize immediately. We might imagine all sorts of systems by which populations A and B would compromise when they are mingled. However, any reasonable system would have to satisfy three conditions:

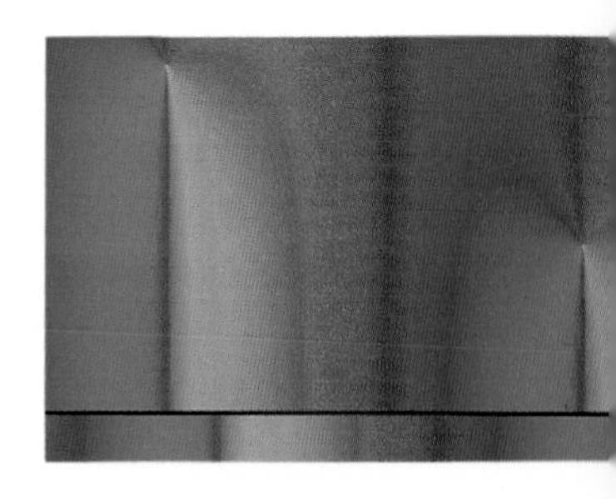

Rule 1 If the initial phase is nearly identical in both volumes, then the compromise phase should be nearly the same as both initial phases. (This virtual separation and recombining happens every instant in an undisturbed volume.)

Rule 2 If the initial phases differ but the two volumes are equal, it should make no difference whether A is ahead of B or vice versa; were we to rename the two populations A and B in the opposite order, the outcome should still be the same.

Rule 3 Slightly changing the phase of either population when they are commingled (the initial phase of A or the initial phase of B) should only slightly change the compromise phase; otherwise, no consistent results could be established in the laboratory.

Is this enough for some interesting conclusion? Yes, in a surprising way: These innocent-sounding requirements turn out to be mutually contradictory. At least one must be violated, at least in some special situation. It turns out to be Rule 3, the seemingly innocuous assumption of continuity. Continuity breaks down to disclose an unavoidable phase singularity. To see how and why, we need only express the rules as an appropriate geometrical diagram and consider how to color it.

The Diagram

We need to display every possible combination of timings when the two populations come together: the phase of A and the phase of B. The phases of A come one after the other, repeating cyclically, so they may be represented on a circle. Perpendicular to that circle and attached to each point on it is a circle of all the possible phases of B. A circle decorated with attached circles makes the surface of an inner tube, or torus. Note that this is not the same torus as in Chapter 3. That one was a framework for depicting the dependence of one phase on another. Here each point on the torus surface represents a combination of two independent phases upon which a third will depend. The third will be indicated by color.

Each point on the torus represents the timing of one mixture, a combination of initial phase A (measured along the horizontal circular axis from some arbitrary zero) with independent initial phase B (measured perpendicularly around the vertical circular axis from the same origin). This diagram represents every possible combination while placing nearly identical mixtures next to each other. For each phase A, mixtures

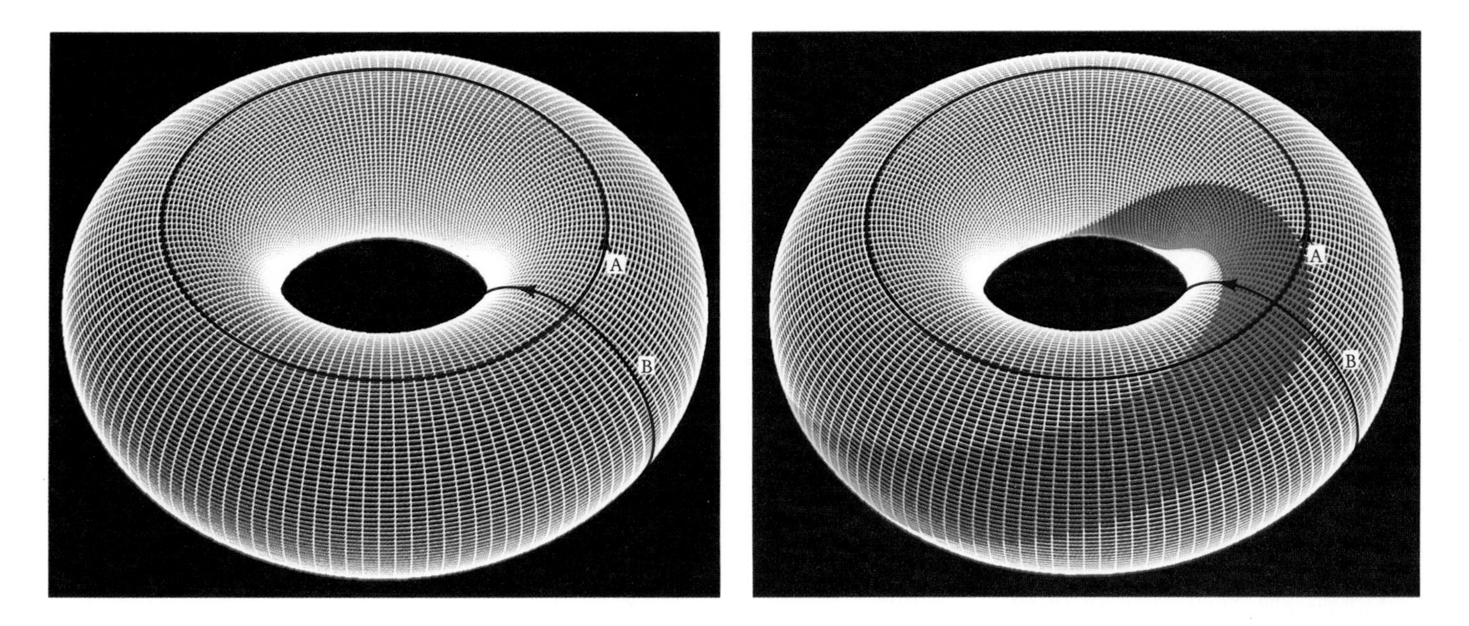

Left: All possible combinations of phase A with phase B can be represented as points on the surface of a torus. *Right:* The resulting compromise phase C after each (A, B) trial can be depicted by color coding the (A, B) point on the surface of the torus. Along the A = B diagonal, the result C = B = A is known *a priori* (Rule 1) and can be color coded as a complete color wheel (by definition of the code) even before measurements are taken.

in which A is held fixed while tested against every phase B lie on a circle parallel to the vertical B axis. For each phase B, the mixtures in which B is held fixed while tested against every phase A lie on a circle parallel to the horizontal A axis. The mixtures in which A and B are identical, whatever their phase, constitute the diagonal line that passes through the intersection of axes and links once through the hole in the torus. Different positions along that diagonal are different phases in the cycle when the two identical populations are mixed (without effect, since there is no way they could know anything has changed). Any other diagonal parallel to it consists of all the experiments in which A and B differ by a given phase shift, but are commingled at different phases scanning a full cycle.

The Colors

To represent the outcomes of all possible mixtures, paint each point on the torus with a color that represents the compromise phase C. That is easy along the A = B diagonal circle, because C = B = A by Rule 1: assign phase zero the color red, as usual, and just paint the full color wheel along that circle. What about the rest of the torus? By Rule 2, the color field is to be symmetrically extended, since an (A, B) experiment is the same as a (B, A) experiment. By Rule 3, every point is to be painted, and neighboring points are to have similar colors. Try it. You can't. And neither can yeast cells.

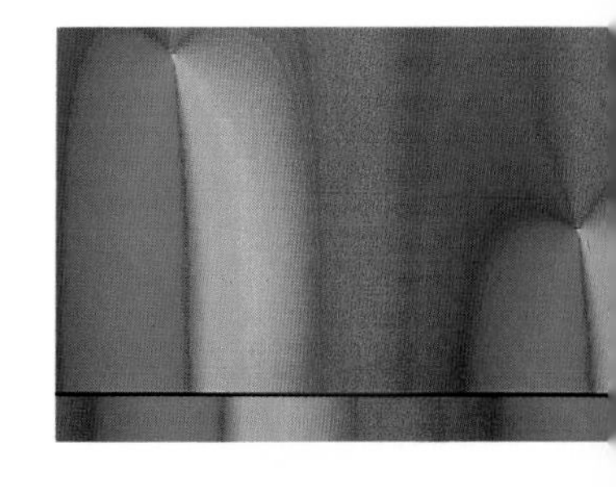

MUTUAL COUPLING AND MUTUAL ENTRAINMENT

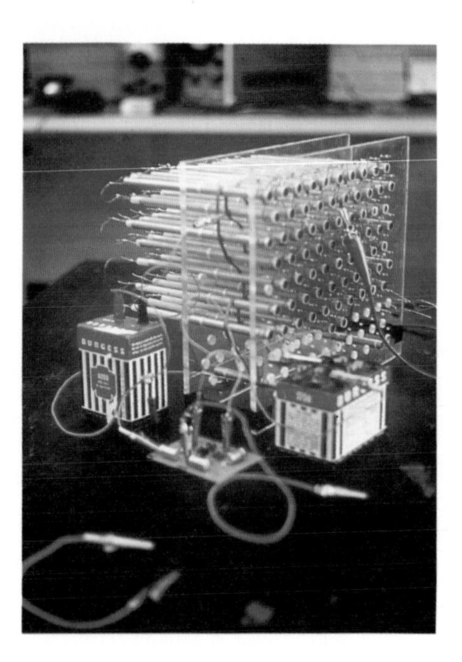

Decades ago, the complex rhythmic patterns of sleep and waking in small mammals were found to be analogous to those observed in populations of mutually coupled electronic oscillators. Seventy-one neon tubes were used in this vintage 1965 model.[6]

With few exceptions, living things on this planet are built of units, in each of which a copy of the organism's genetic code dominates chemical activities in a tiny surrounding volume. These cells are units of function, not just of structure. In *Gonyaulax*, cellular functions include circadian timekeeping. The many cells in a population can interact weakly like mutually synchronizing clocks. Twenty years ago, biologists already suspected that such interactions could be important for maintaining tight coherence and enhancing individual regularity, and even in some cases for ensuring *asynchrony* despite the environment's rhythmic influence. At that time, such consequences of coupling between chemical clocks could only be foreseen by building electronic caricatures, by computational simulation, and by mathematical analysis.[5]

Today, biologists are almost able to check those anticipations by actual experiments with populations of cellular circadian clocks. Four prototypes are already in hand: a strictly nonliving acellular chemical oscillator discussed in Chapter 9; a living acellular oscillator discussed in the box on pages 132–133; the yeast cell suspensions discussed in this chapter (which actually are more intensively studied in a form more like the other two examples, as a cell-free extract); and a neural pacemaker population in the brain of electric fishes. In all cases, the coupling between clocks in the population is much stronger than between *Gonyaulax* cells. In two cases, the coupling intensity can be varied experimentally.

In suspensions of yeast cells, coupling strength can be varied by diluting the suspension. One of the more robust and striking predictions of the theory of mutual synchronization was that it should fail abruptly below a critical coupling strength. John Aldridge and E. Kendall Pye tried this experiment with yeast and found exactly that: when the cells get more than about twenty diameters apart, the amplitude of their collective rhythm falls abruptly.

Similar trials are in progress at the Scripps Institution of Oceanography with one of the most precisely regular biological clocks known, a collection of pacemaker neurons in the brain of the electric fish *Apteronotus leptorhynchus*. The fish uses this population of clocks to drive its electric organ at audio frequencies for navigational purposes in dark water; but scientists Walter Heiligenberg and John Dye can use it to study the dynamics of mutual synchronization. The pacemaker nucleus of the fish brain, being made of neurons, is a closer analog than suspensions of yeast cells to the suprachiasmatic nucleus of the mammalian brain, the putative center of circadian rhythmicity in man. After flooding the nucleus with possible neurotransmitters, Dye observes that its cellular oscillators, no longer able to communicate, quickly lose both their former mutual synchrony and their incredible precision.

James Enright, also of Scripps, has shown theoretically by computer simulations that a population of nervelike elements, even if separately unable to function like clocks, can nonetheless collectively constitute a pacemaker of extraordinary regularity. Even though in his hypothetical nerve network an isolated cell follows none of the rules of clock resetting explored in this book, still the collective of cells turns out to have even and odd resetting, a time crystal with screw surface, and a phase singularity.[7]

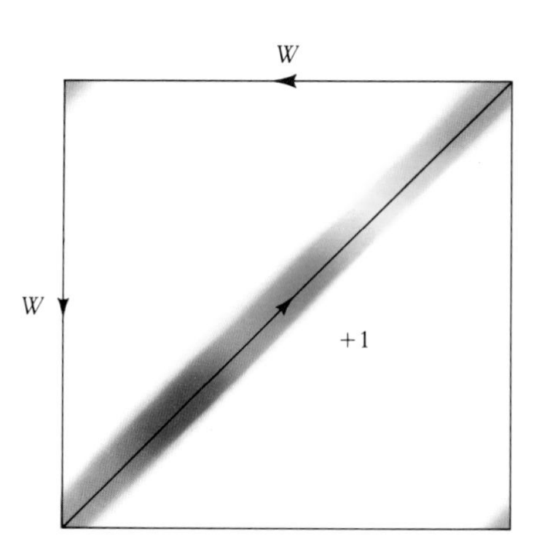

The skin of the torus can be laid flat like a rug, revealing a symmetric pair of triangles joined along the diagonal backbone. The winding number of phase C along the diagonal is 1 and is an integer *W* on each of the remaining sides. Positive *W* connotes an increase of C as A or B increases.

To save yourself the expense of a thousand good inner tubes and a clean shirt, consider the problem this way. With a sharp razor, slit open an inner tube along the circular A axis then along the circular B axis. Flatten the colored rubber rectangle onto your desk top. Think about the colors along the triangular path indicated in the illustration on this page. Follow it in a counterclockwise direction, starting with an ascent to the right along the diagonal: the colors run forward through one full cycle, starting at red and ending at red. Next, as A increases one cycle along the roof of the triangle the colors must go through some exact integer number of full cycles; call it *W* for *winding number*. Why integer? Because the two end points of this side are the same color (red).

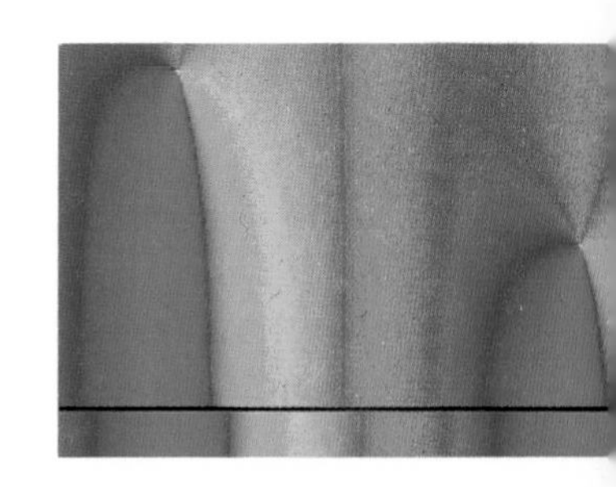

(Remember, each edge of this square was a circle on the intact inner tube: its end points are the same point, cut in half.) Going through the color wheel from red to red we must have traversed an exact integer number (including zero as a possibility) of complete cycles. The story is the same while continuing down the left side of the triangle, because it is the left side of the rectangle, which is the same as the right side, which, on account of the symmetry of A and B, is the same as the roof traversed right to left. So the same integer number of cycles of color (W) is passed through again.

How many cycles all the way around? $1 - W - W$. (The negative of W is taken because we are moving counterclockwise, decreasing A or B along the rectangular edges.) In the simplest case, W would be 0 (the edges of the rectangle are all red) and circumnavigation takes us through exactly one cycle. Déjà vu again? It sounds like the missing day in Pigafetta's logbook. To pursue this thought experiment to its logical conclusion, next imagine some negligible distortion of this neat triangular path. Since the colors change smoothly according to Rule 3, they won't be much different along the new path. In particular, there is no way a complete cycle of color could suddenly get left out or added in. So the winding integer persists unaltered. Then shrink the path a little. Still, the number of cycles can't jump by a full cycle. Continue: shrink the path more and more, ultimately to a very tiny ring, scarcely varying in its A and B coordinates. How many full cycles of color lie around that tiny ring? Still $1 - 2W$. Perhaps you expected 0; after all, according to Rule 3, the result shouldn't change much when the initial conditions scarcely change. However, no integer W will make $1 - 2W = 0$; zero is even, and $2W$ is even, but $1 - 2W$ is inescapably odd.

The Paradox

It seems that there is no coloring compatible with Rules 1, 2, and 3. The best we can do is to say that those innocuous rules must be respected everywhere except at some isolated point (at least one) in each triangular half of the torus-skin. For example, suppose all the borders of the rectangle are yellow, which means that when one population is at that phase, it will completely dominate the other population, whatever its phase may have been. Then $W = 0$ and $1 - 2W = 1$, so a single cycle of color converges to the unavoidably exceptional point. The point is a phase singularity. It has no definite hue, as it shares equally in all the hues represented in our palette of coding colors; it is some kind of average of all, perhaps gray—in any case, hueless.

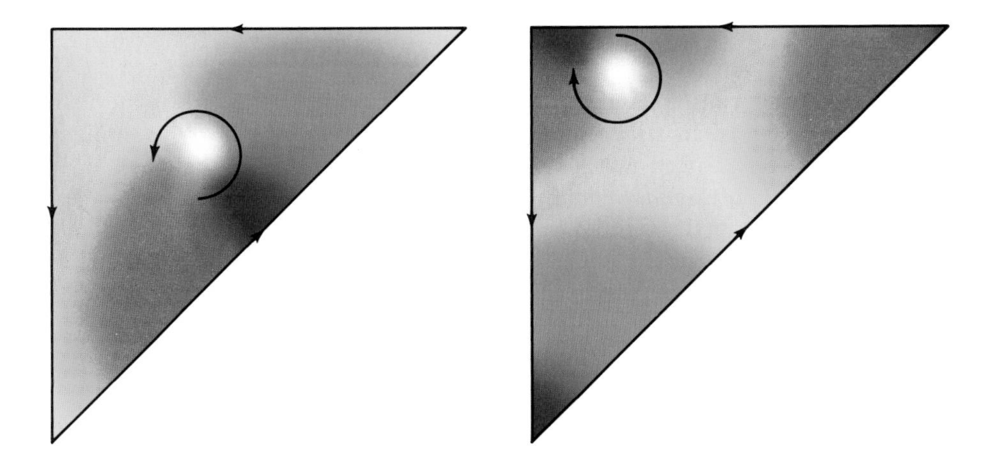

Left: If $W = 0$ along each edge of the rectangle like the one on page 129, then compromise phase C increases *counterclockwise* around the singularity. *Right:* If $W = 1$ along each edge, then compromise phase C increases *clockwise* around the singularity. The colored data plot on page 135 shows that both cases occur in suspensions of yeast cells.

Consider another extreme possibility: when one population is at the purple phase, it is utterly susceptible to domination by the other, regardless of phase. In this "recessive" case, the colors progress through one cycle along each border of the rectangle; that is, $W = 1$. In this case, $1 - 2W = -1$. Again the singularity consists of a single complete cycle of hues converging at the singular point in each triangle—but this time their ordering is reversed.

These two extreme cases have special significance in terms of what the preceding chapters revealed about phase resetting. From the viewpoint of population A, receiving a dose of population B is like exposure to a stimulus. If B is always added when it is at the same phase, so it is a constant stimulus, but it is added at different times in A's cycle, then a plot of the compromise phase C against old phase A is like a resetting curve. We have seen that those curves come only in two styles: odd, with winding number $W = 1$, and even, with winding number $W = 0$. It seems reasonable to watch for these in the yeast experiment.

Other examples can be imagined in which W has values other than 0 or 1. In such cases, two or more such singularities will be required in each triangle. The only way there could be no singularity is for W to be ½, which means either that the color of a corner of the rectangle differs by a half cycle from its color in a replica at either adjacent corner (a plain violation of logic, if we insist on continuity) or that there is a half-cycle discontinuity in the results, a crack in the diagram that must cut twice across every version of our path from the top of the diagonal back to the bottom. (If you think this last is crazy, see the box on the next page.)

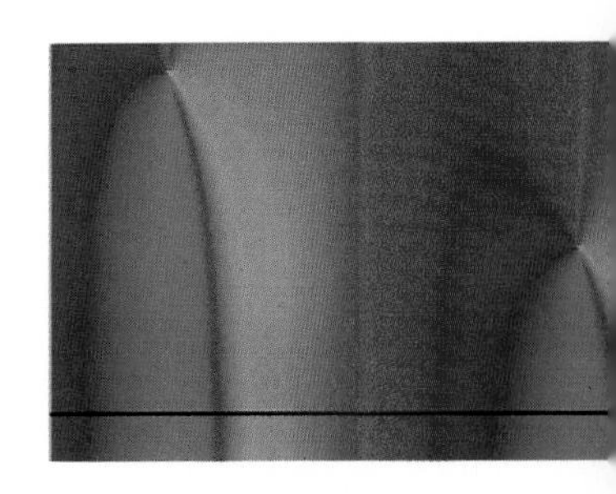

NUCLEAR FISSION IN SLIME WITHOUT CELLS

The yellow slime mold *Physarum* pulses as it spreads over a rock on a Pennsylvania forest floor.

Most living organisms are built of cells because one nuclear copy of the genetic code can efficiently dominate only a limited volume of the reacting gel around it. But it need not always be the *same* volume. In the true slime mold *Physarum polycephalum*, the usual partitioning of cytoplasm into cellular parcels, each with its unique private nucleus, is abandoned in favor of anarchic churning of all the nuclei throughout all the cytoplasm. Such an organism cannot grow by cell division: instead it just expands while nuclei replicate from time to time as required lest they grow progressively more dilute. Because they coinhabit one big cell, all nuclei simultaneously sense that it is time to untangle the DNA and divide: the event takes about 15 minutes in a cycle of 10 hours or so.

How is this mutual synchronization arranged? Hints were first obtained by mixing together two such blobs of jelly, A and B, each synchronous within itself, but each set for nuclear fission at a different time. What compromise phase C was established, and how did it depend on the volume ratio and the phases A and B at the moment of fusion? A decade ago, systematic experiments of this sort were set up to test the principles of phase compromise as summarized in this chapter.[8] The results scarcely resemble those obtained with yeast. They are instead dominated by conspicuous discontinuities like earthquake fault lines: an hour or two before nuclear fission would have occurred in either A or B, the compromise phase C changes abruptly by an amount proportional to the ratio of the volumes of A and B. No phase singularities are evident: in the region outside those two two-hour-wide bands the compromise is just the average of the two parent phases. These observations are widely taken to favor the idea that some chemical accumulates between nuclear divisions until a critical concentration is reached, at which point all nuclei become irreversibly committed to divide, and the accumulated excess chemical vanishes. Phase compromise represents the volume-weighted averaging of two concentrations.

It is possible that the mechanisms of nuclear fission are that simple; they might also be more subtle, the evidence being hidden in the considerable area occupied by ostensible discontinuities. But if we take the clear part of the evidence at face value and extrapolate it to fill the resetting diagram except along hairline cracks, then there is a take-home lesson: not every biological clock abides by the topological principles first discovered in circadian clocks.

In slime molds, the fundamental cause of the difference is discontinuity. But in "A Parable of Three Clocks," in Chapter 3, a different case was shown, in which there were no discontinuities of mechanism but, anyway, no even resetting or singularity. It is time to ask what kinds of clock are resettable in the ways depicted by a time crystal with phase singularities. Chapter 8 takes up that question.

In short: barring cracklike discontinuities in the dependence of compromise phase upon the two parent phases, the mere logic of the experimental design forces its results to contain at least a point discontinuity. In that case, it has to be the most violent kind—a phase singularity.

A cynic might conclude from this thought experiment that biological theory belongs in a book by Lewis Carroll. Here again it has come up with nonsense equivalent to those paradoxical "proofs" that $1 = -1$ that students encounter while studying algebra. The cynic would be wrong. There is an example of Rules 1, 2, and 3 in more familiar experience. Suppose we are combining not chemical clocks coded by color, but colors themselves. Suppose we shine two flashlights, each covered by a colored transparency, onto a white wall, and ask what hue we may expect to see. The rules are the same: switching transparencies between the two flashlights makes no difference, hue on same hue makes same hue—and sufficiently slight changes of either flashlight's color should result in only slight changes of the color seen in the overlap region. To do the whole experiment in one shot, make a striped transparency that runs through at least one full cycle of hues, and place it in a projector with stripes horizontal to establish the vertical phase direction on our wall. Place another piece with stripes vertical in another projector to establish the horizontal phase direction, and overlap the first projection. What do you get? A colorful square containing all combinations of hue with hue, a full cycle of hues running up the diagonal that divides the square into two mirror-image color patterns, and—a gray spot in each triangle!

The Experiment

What will yeast cells actually do when combined in equal volumes at that unique combination of moments? God may be a mathematician, as theorists are fond of saying, but cells are biochemists. They abide by the rules of chemistry, not necessarily by the rules of simpler mathematics. (Perhaps God is a subtle mathematician.) Cells can get out of these

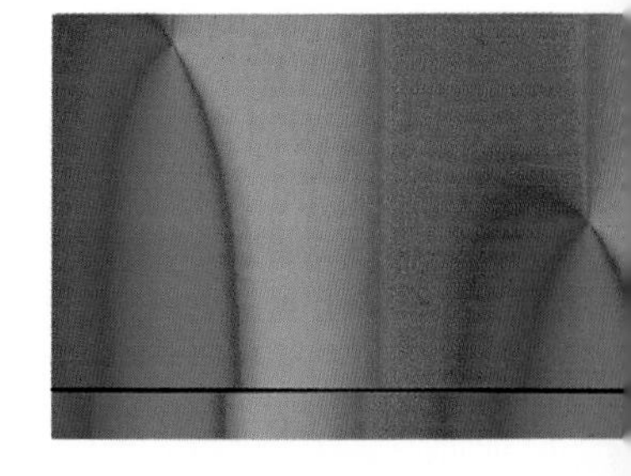

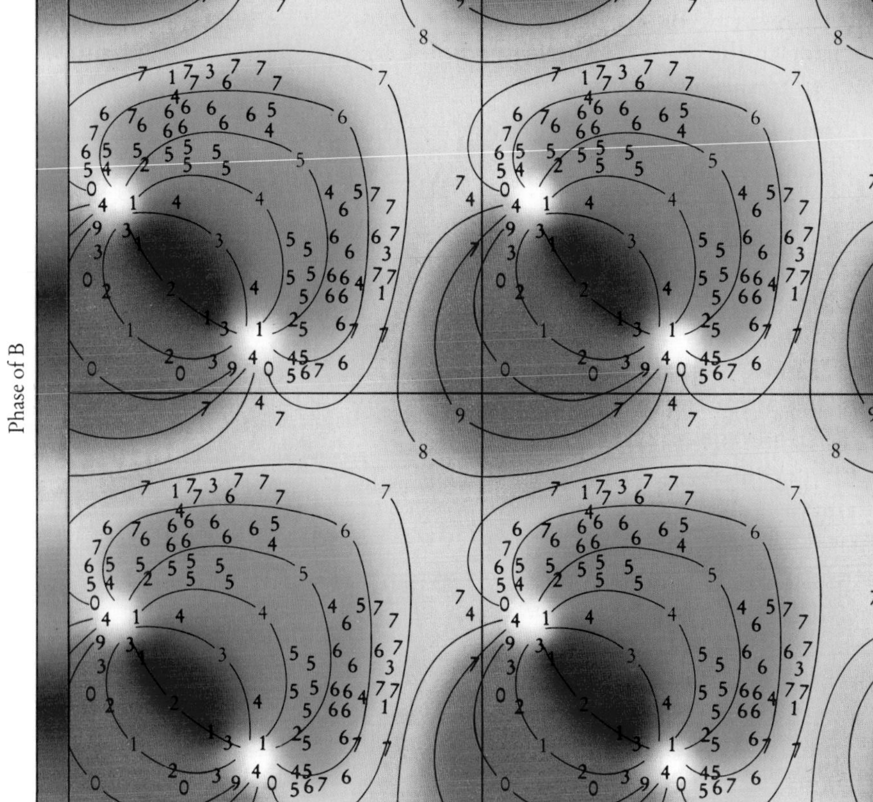

Left: Color-coded experimental phase compromise data are plotted here along two cycles in each direction. Phase zero (red, at the edges of each square) here represents a moment 1/20 cycle before NADH fluorescence maximum: the maximum occupies the first 1/10 cycle after each edge, as denoted by the "0" and subsequent decimal digits along the diagonal. Contour lines follow uniform-color "isochromes" from one singularity to another. The digits encode results from the yeast mixing experiment in tenths of a cycle. *Right:* The colors are repeated without contour lines or data to show that the unit cell can be defined in any position. The two particular choices shown correspond to the extreme cases foreseen in the triangles on page 131.

mathematical dilemmas by even more flagrantly violating Rule 3: complex machinery can have all sorts of discontinuities in its behavior. Some kinds do (see the box on page 132), but the yeast cells seem not to. They make their phase compromises just as described above.

A. K. Ghosh, Britton Chance, and E. Kendall Pye in the University of Pennsylvania School of Medicine[9] did the experiment in 1970 with a clean mind, unbiased by a theory of phase singularities, which had not yet been applied to this case. As in all experiments, the results were not perfectly consistent; as in many, their pattern was not evident even to shrewd investigators who knew the inner workings of yeast cells intimately and had expectations on that basis. But the pattern that seems to

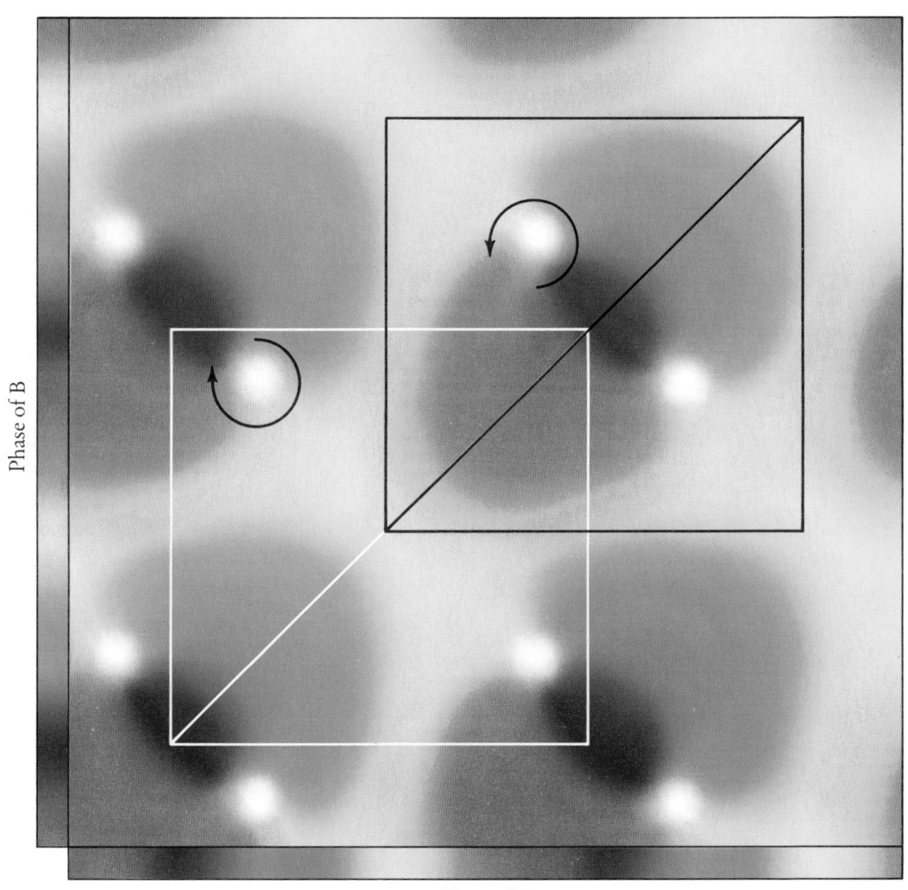

fit best is shown on this page as four sewn-together copies of the colored inner tube. It satisfies Rules 1, 2, and 3 except at two symmetrically placed singularities.

Notice that both triangles shown on page 131 appear in these experimental results. The facing page shows a black unit cell burned in atop the color field, with its edges (phase zero for both A and B) displaced to what was originally phase ¾. This is a "dominant" phase, at which $W = 0$. The edges of this box show even resetting: the compromise phase color is nearly the same all around. The singularity in each triangle therefore turns with the progression of old phase along the diagonal. The white unit cell has edges (phase zero for both A and B) displaced to

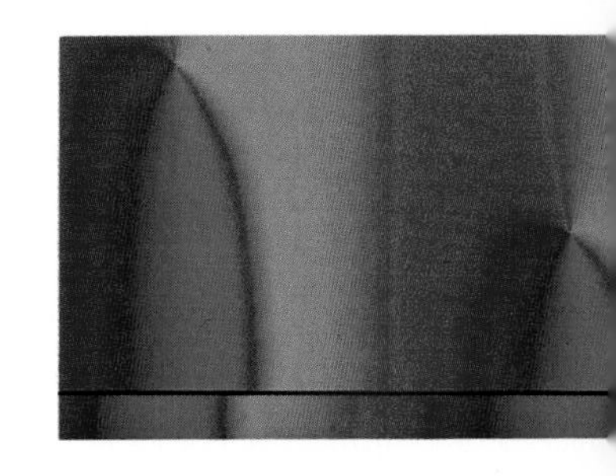

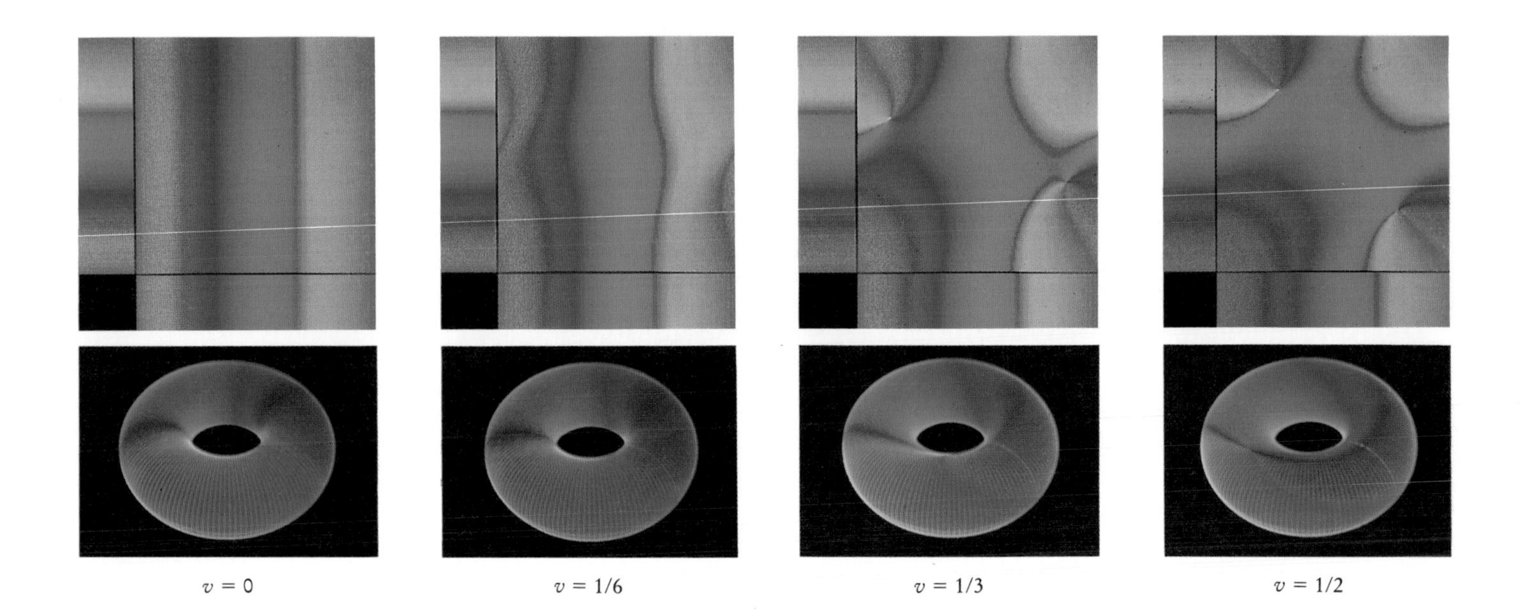

Top: Phase-compromise planes are painted for several values of v. Level v = ½ corresponds to the square outlined in black on page 138.[11] Phase zero is chosen here to frame the contours as in the "recessive" framing of the right half of the figure on pages 131 and 135. In each panel, phase A increases from 0 to 1 left to right, and B from bottom to top. *Bottom:* Each of the top seven squares represents the surface of a torus, peeled off and flattened. On the torus, phase A increases along the horizontal ring axis, to the right in the foreground. Phase B increases, upward in the foreground, along the vertical ring axis. The coordinate origin (A = B = 0 = 1 = red) is in the lower right of each panel.

what was originally phase ¼. This is a "recessive" phase, at which W = 1. The edges of this box show odd resetting. The singularity in each triangle therefore turns opposite to the progression of old phase along the diagonal.

What happens at the singularities? The system won't stay at ambiguous phase after the critical stimulus, but neither is it allowed to have any predictable phase, so it becomes unpredictable. The biochemists wrung their hands over the irreproducibility of their results in certain regions: "Synchronization does not follow any clear pattern. The reason . . . is not understood at this time. Much more experimentation is probably needed."[10] But no one could *possibly* have obtained precise phase measurements throughout this experiment. The mathematical prediction seems to be that you can hire any number of theorists—or even experimentalists, for that matter—to tell you the compromise phase and none will agree (except those who stay silent). Ernest Rutherford, the great experimentalist who discovered the atomic nucleus, once remarked against theorists, "They play games with their symbols, but we turn out the real facts of nature." What would he have said in a case like this, where the "real fact" is that an unambiguous result simply cannot exist and so cannot be obtained even in principle? The ostensible conflict

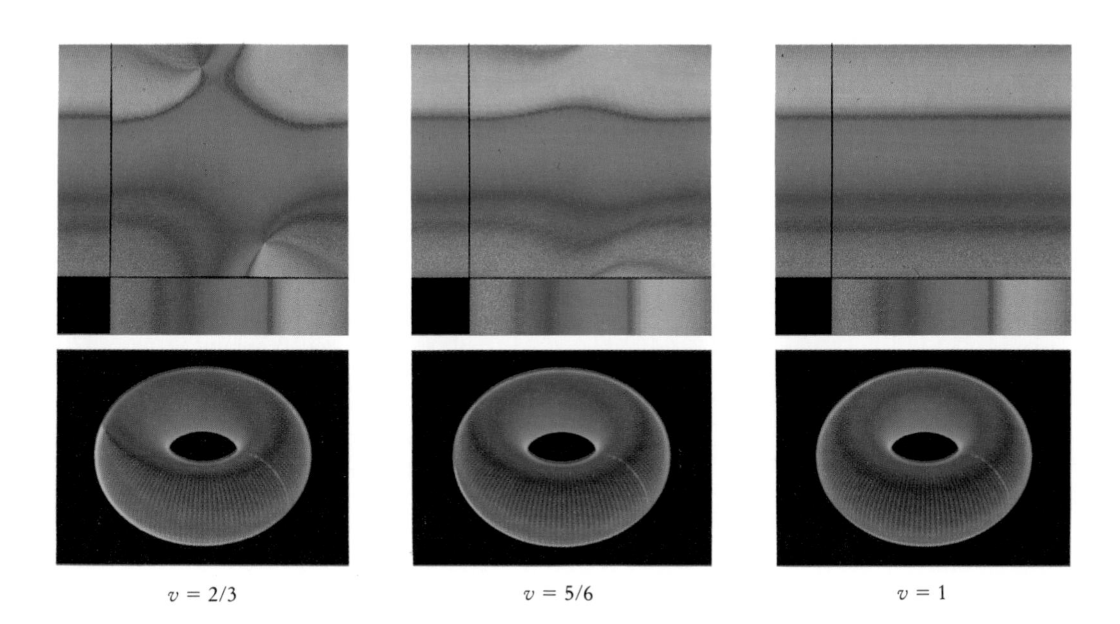

$v = 2/3$ $v = 5/6$ $v = 1$

between theory and experiment is a pernicious illusion: neither provides an interpretation of the world without the other.

The Rest of the Story

In the arbitrary symmetry of the mixing experiment, we used exactly equal volumes, and the logic of our inference depended explicitly on that condition. But the whole story of phase compromise should include the results of asymmetric mixes, too. For each volume v of A and complementary volume $(1 - v)$ of B, there is a distinct square like the one on pages 134 and 135. We already know how these squares are colored in the limiting cases $v = 0$ (pure B) and $v = 1$ (pure A): compromise phase C is unaffected by the phase of A at volume $v = 0$, and is unaffected by the phase of B at volume $v = 1$. Intermediate cases, calculated from a simple analogy to the kinetics of oscillating glycolysis,[12] are shown above.

Assembling the entire stack of colored squares like the floors of a building (figure on the next page) reveals the color pattern on the four side walls of the cube. On the south face are experiments in which volume v of A is tested at each phase in its cycle by adding a complementary volume $(1 - v)$ of B at one chosen phase. This complementary

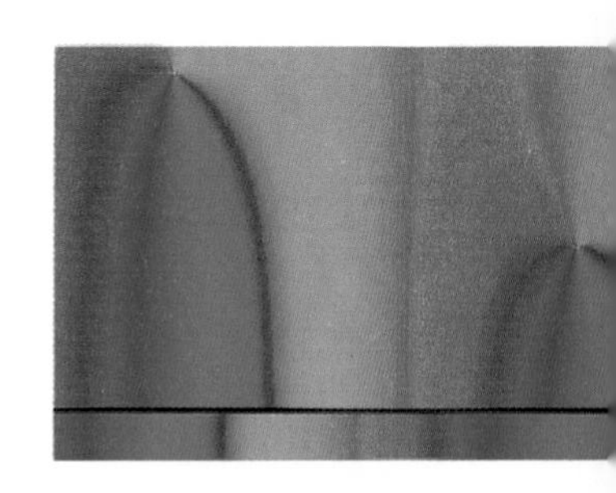

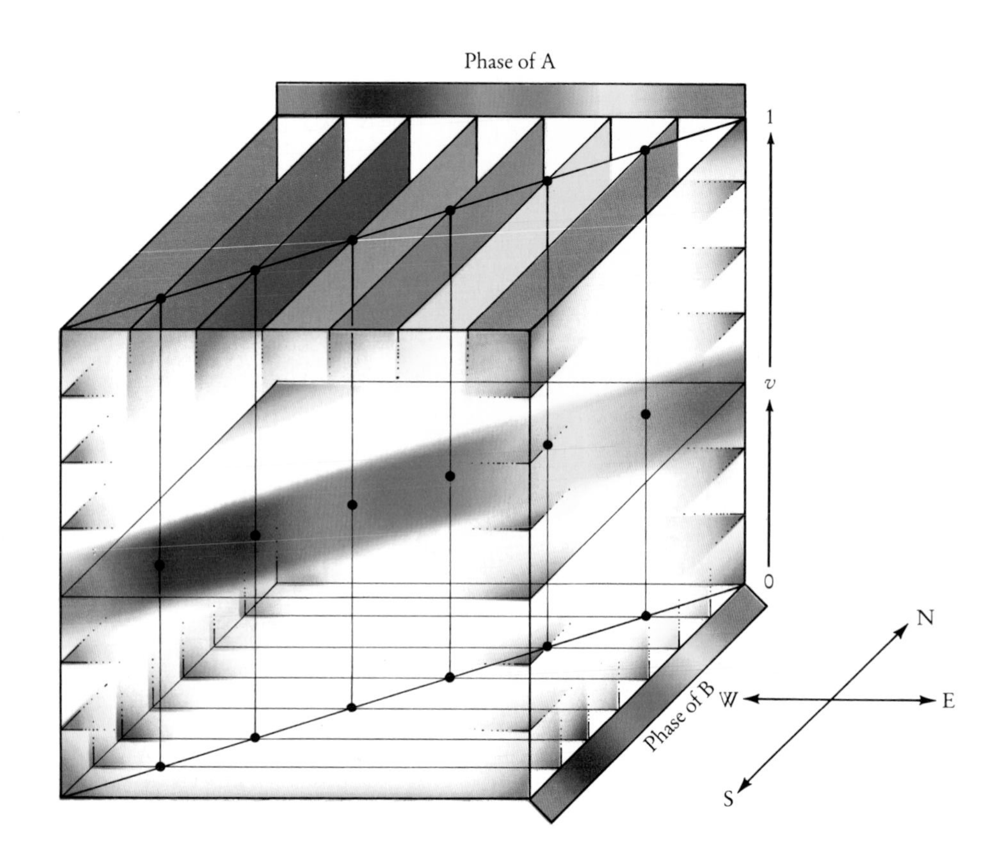

One unit cell of the preceding illustrations is placed in a larger context as the horizontal square lying at level $v = 1/2$ in a cube that spans the range of A volumes 0 to 1 (B implicitly has the complementary volume $1 - v$). At $v = 1$ the phase "compromise" is independent of B (since none is added), and at $v = 0$ it is independent of A. Along the diagonal plane where the phases of A and B are equal, phase C increases through one full cycle and those results are independent of the volume ratio. Results are symmetric about this plane only when $v = 1/2$. Surfaces of constant phase (color) fill this cube. Their edges are indicated near $v = 0$ and $v = 1$.

volume is like a standard stimulus applied at different times in the cycle, in different amounts: have we seen this protocol before? This is nearly the format of all the phase-resetting experiments described earlier, so it is not too surprising to see that this face of the cube turns out to be colored in a similar pattern, as in the figure on the facing page.

Next, we know that the north-south pair of opposite faces are identical because they represent the same phase of B, only one cycle apart; and the east-west pair of faces, at right angles to those, represent the same experiments with A and B interchanged, so v interchanges with $(1 - v)$ and the faces are upside-down in relation to the first pair.

Finally, what about the interior? There was nothing special about the phase we chose for the nominal "ends" of the cycle, exposed on four faces. So we expect interior "faces" parallel to those to look qualitatively the same. They do. In fact, the phase singularity weaves sinusoidally through the interior, emerging from one face as it reappears at the

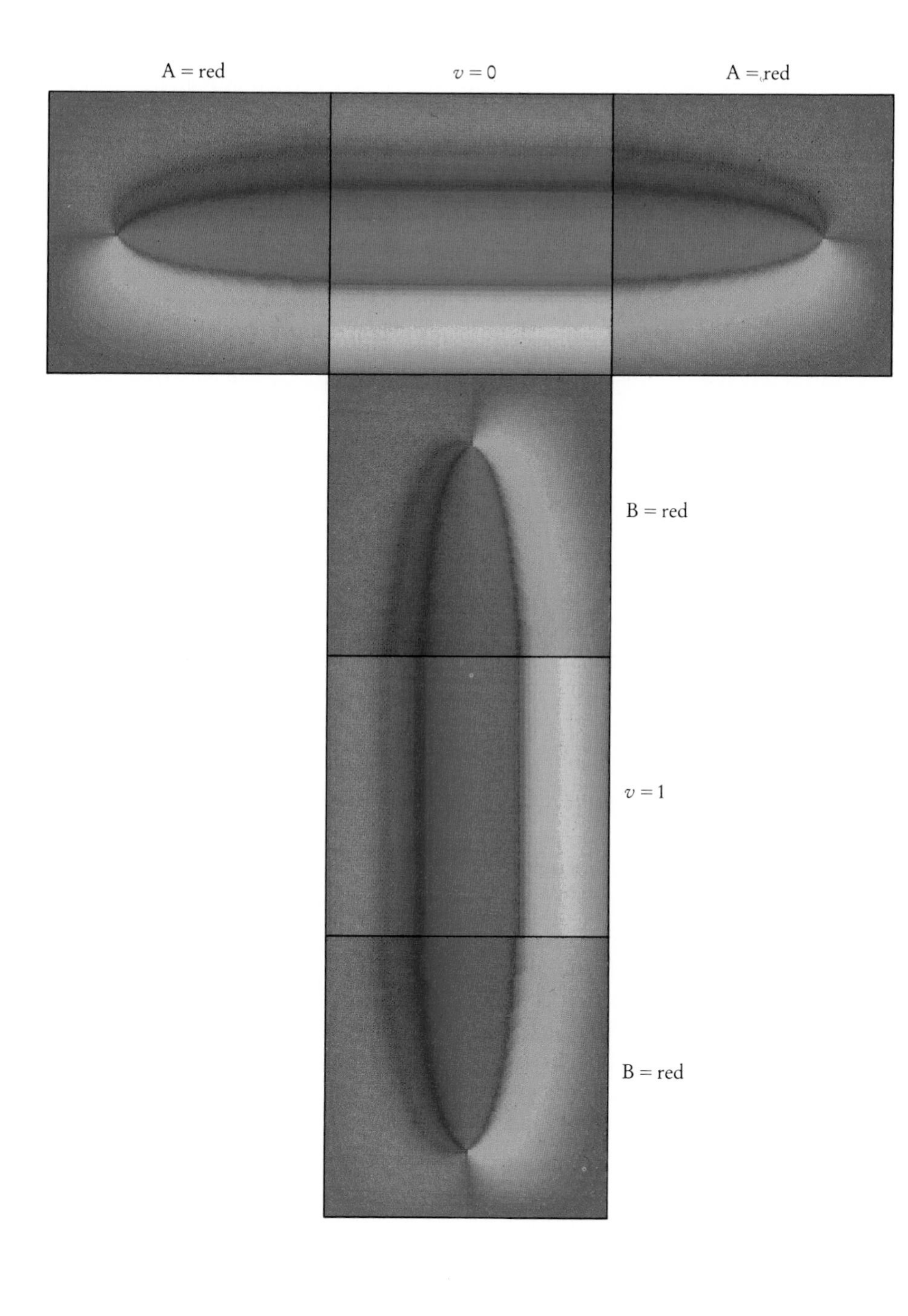

The six faces of the cube shown on page 138 include panels $v = 0$ and $v = 1$ from pages 136 and 137 and four identical copies of a contour map like the one on page 125.

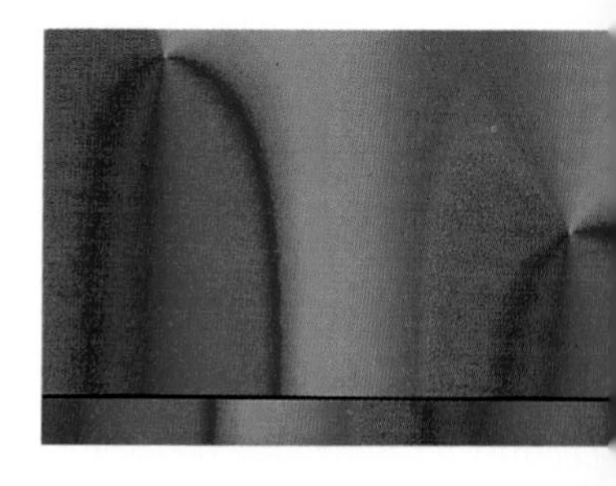

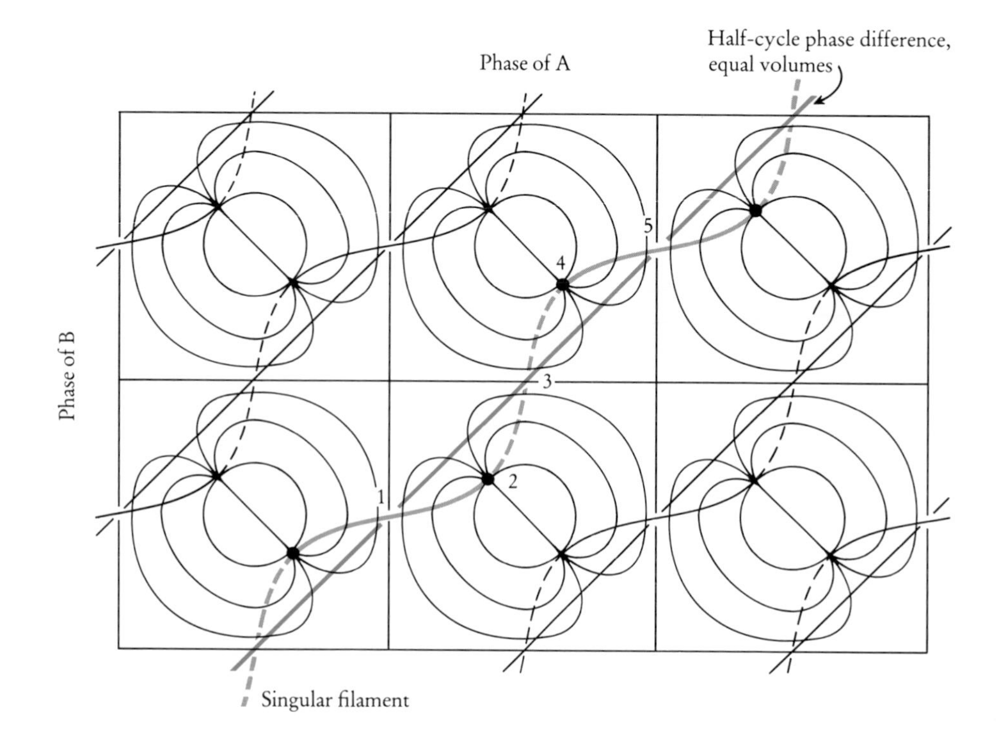

Several identical cubes like the one on page 138, but with dominant framing are repeated here in both directions; only the contours of the midpoint planes $v = ½$ are shown. Imagine $v = 0$ above the paper plane and $v = 1$ below. The singularity threads its way (1-2-3-4-5) through the cubes between the emergence points shown in the preceding illustration. The isochrome surfaces of uniform phase C all converge to this filament of singularity, depending from it like the blades of a turbine (as seen in cross section on four of the six faces in the preceding illustration.)

opposite face. In this fuller three-dimensional picture the singularity is a one-dimensional filament, not the mere point that appeared on every two-dimensional cross section. The singularity in the vertical pinwheel-like faces is the same one seen twice in the horizontal phase-compromise diagrams! In the picture above, follow the filament from its entry at (1) through one side wall of the cube, up through one triangle in the plane $v = ½$, out through the back wall into the next unit cell, down through the other triangle of the plane $v = ½$ and finally back to its entry point (displaced by 1 cycle in both A and B directions) at (5). It has traced one coil of a helix around the diagonal that represents a constant phase difference of one-half cycle between A and B in the plane $v = ½$. Meanwhile, the color field centered on this singularity has also rotated through one full turn.

The Mechanisms of Circadian Clocks

The experimental study of biochemical timing in the one-minute oscillations of sugar metabolisms has already consumed decades of effort by innumerable dedicated investigators; it would take much longer to execute a comparable analysis of 24-hour oscillations. Nonetheless, parts are already underway. Earlier chapters described phase resetting experiments and the resulting singularities and time crystals in a diversity of circadian systems. Other experiments were designed to detect phase compromise through interactions between cells with differently set clocks. Weak mutually synchronizing interactions have been observed in *Gonyaulax*, and much quicker mutual synchronization is known or suspected between the symmetrical halves of the brain in several kinds of insects, snails, reptiles, and mammals. Unfortunately, almost all this work is still limited to the final effects on some behavioral rhythm. It would help to know more about the mechanisms underlying the circadian time sense.

In some cases, the mechanism is biochemically trivial, being fundamentally external; a dead horse, for example, still has diurnal ups and downs of body temperature due to the rising and setting of the sun. Franz Halberg's term *circadian* was adopted a quarter century ago to distinguish the more interesting instances of diurnal rhythmicity in which the mechanism is fundamentally internal. In such cases, the period depends on internal factors—temperature, genetic background, percentage of protons replaced by deuterons, and so on. Unless it is entrained by an external pacemaker, the free-running period is only *circa* (roughly) "dian." The persistent discrepancy of the internal period from that of external cues results in arbitrary phase differences between the circadian rhythm and local time. Those differences can be altered by a phase-shifting stimulus, and the alteration is stable. These classical observations indicate that the mechanism is not dependent on external pacemakers but has an internal, and fundamentally biochemical, mechanism.

It has proven difficult to infer anything about molecular mechanisms by combining insights obtained from different organisms. The basic reason might be that the circadian mechanism is not universal, but differs from one species to the next, or even from one cell type to the next within a species. But there is much reason to believe that the mechanism(s) is complete at the cellular scale, unlike the menstrual cycle, for example, which involves the hormonal or electrical interplay of diverse tissues. The cellular mechanisms can be probed biochemically and by techniques of genetic engineering. There are many ways to interfere biochemically with the workings of the circadian clock. The earliest approaches emphasized the universal sensitivity to light. In *Drosophila*,

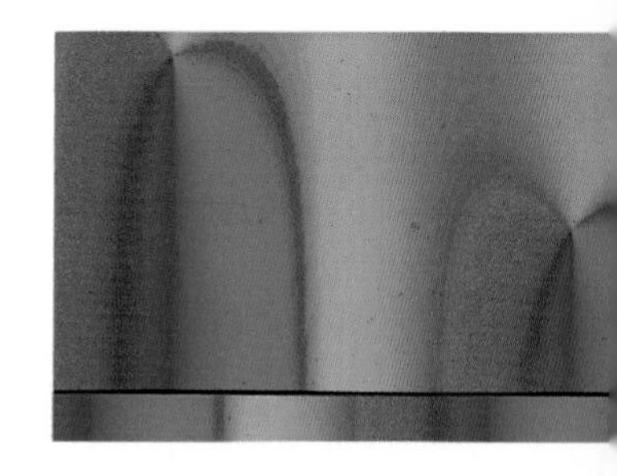

for example, steady illumination even at the level of partial moonlight is sufficient to paralyze the clock. But in every species examined up to now, it appears that the impact of light is indirect, not directly on a molecular component of the oscillating mechanism. Hints were also taken from the fact that most circadian rhythms have nearly the same period at any constant temperature in the physiological range. However, it turns out that temperature-independence of period is not unusual in chemical kinetics, and is still less unusual in physiological regulatory systems.[14] Moreover, circadian clocks lack the more revealing property of overall temperature independence; even slight changes of temperature can be powerful phase resetters. So not much about the mechanism has been learned through the two normal channels by which environmental cues regulate it.

In addition to temperature changes and photons, many chemical agents that alter membrane permeability or disrupt protein synthesis prove to alter the period of cycling during chronic exposure. Applied briefly, then allowed to decay, they cause a phase shift. But the processes affected are many and diverse, and no one yet knows how their impact on the clock process may be mediated. It seems that neither ATP itself nor the process of ATP synthesis and degradation are part of the clock mechanism; nor does protein synthesis demonstrably constitute a required step in the circadian cycle, at least not in the breadmold *Neurospora*.[15]

Another approach makes use of mutants in which some aspect of clock function is permanently affected. The simplest aspect to monitor is, of course, the period. Mutants at the *frq* locus of *Neurospora* and at the *per* locus of the fruit fly *Drosophila* have periods ranging from 19 hours to 29 hours, and there are some with no period at all (arrhythmic). Genetic hybrids combining different numbers of mutations adopt a compromise period as though clock rate depends on gene-mediated reaction rates, but there is little hint of the nature of the reaction.[16] It has recently proved feasible to clone the altered gene from *Drosophila*, to sequence it, and to seek homologous sequences in libraries of genes of known function: it appears to be a proteoglycan in the cell membrane.[17] Before this book is published, the role of a clock-affecting gene may be revealed. That might only identify something that has only an indirect effect on the clock process; such was the unsatisfying outcome of decades of effort to identify the photoreceptor pigment in many species. But with luck, the gene product could turn out to participate actively in the daily ups and downs of the circadian mechanism.

One hint to the locus of that mechanism has turned up in *Neurospora* in Stuart Brody's laboratory at the University of California at San Diego. Clock period is changed in a mutant in which a certain molecular fragment of an ATP synthetase in the mitochondrial membrane has been altered.[18] Should this hint prove important, it might account for the absence of circadian rhythms in cells that lack mitochondria.

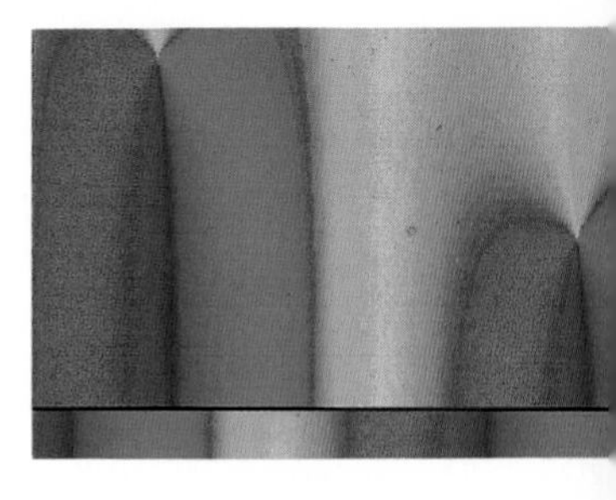

A crescent of dawn has raced around the globe a trillion times while life evolved on this rotating planet.

Chapter Eight

Circadian Dynamics and Its Evolution

Is mathematical analysis . . . only a vain play of the mind? It can give to the [scientist] only a convenient language; is this not a mediocre service, which, strictly speaking, could be done without; and even is it not to be feared that this artificial language may be a veil interposed between reality and the eye of the [scientist]? Far from it; without this language most of the intimate analogies of things would have remained forever unknown to us; and we should forever have been ignorant of the internal harmony of the world, which is . . . the only true objective reality.

HENRI POINCARE

A phase singularity is a point in the interior of a smooth screw-shaped surface that shows how new phase varies with stimulus size and timing. This surface is worth a moment of contemplation, as it is central to the most fundamental capacity of any biological clock: its ability to reset on cue. Unfamiliar as this shape may be, it corresponds to an equally strange but very simple idea: that in some sense something like the external time zones lurk in the inner workings of biological clocks. Deferring interpretation for a moment, consider a metaphor.

Think of the earth, ponderously rotating on its polar axis once in every 24 hours, as a giant clock. As in the figure on the next page, artificially divide the earth's surface (the southern hemisphere, or in fact the Antarctic Continent alone, will suffice) into time zones converging to the South Pole. Now consider equivalent pieces of this clock—clods of earth—situated on the surface in a circle, a parallel of latitude, which passes through all time zones. Each clod is a sample of the clock proc-

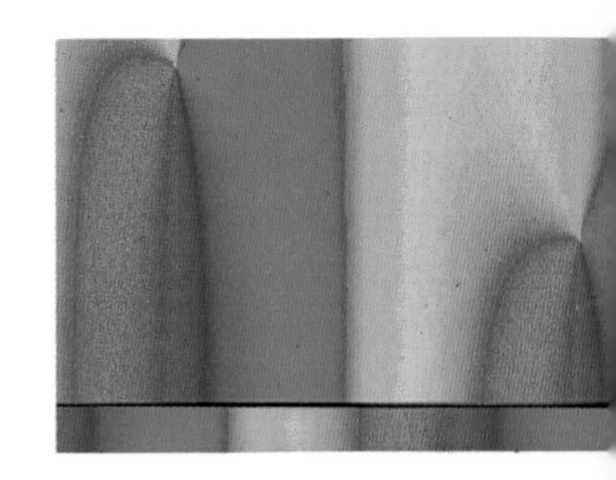

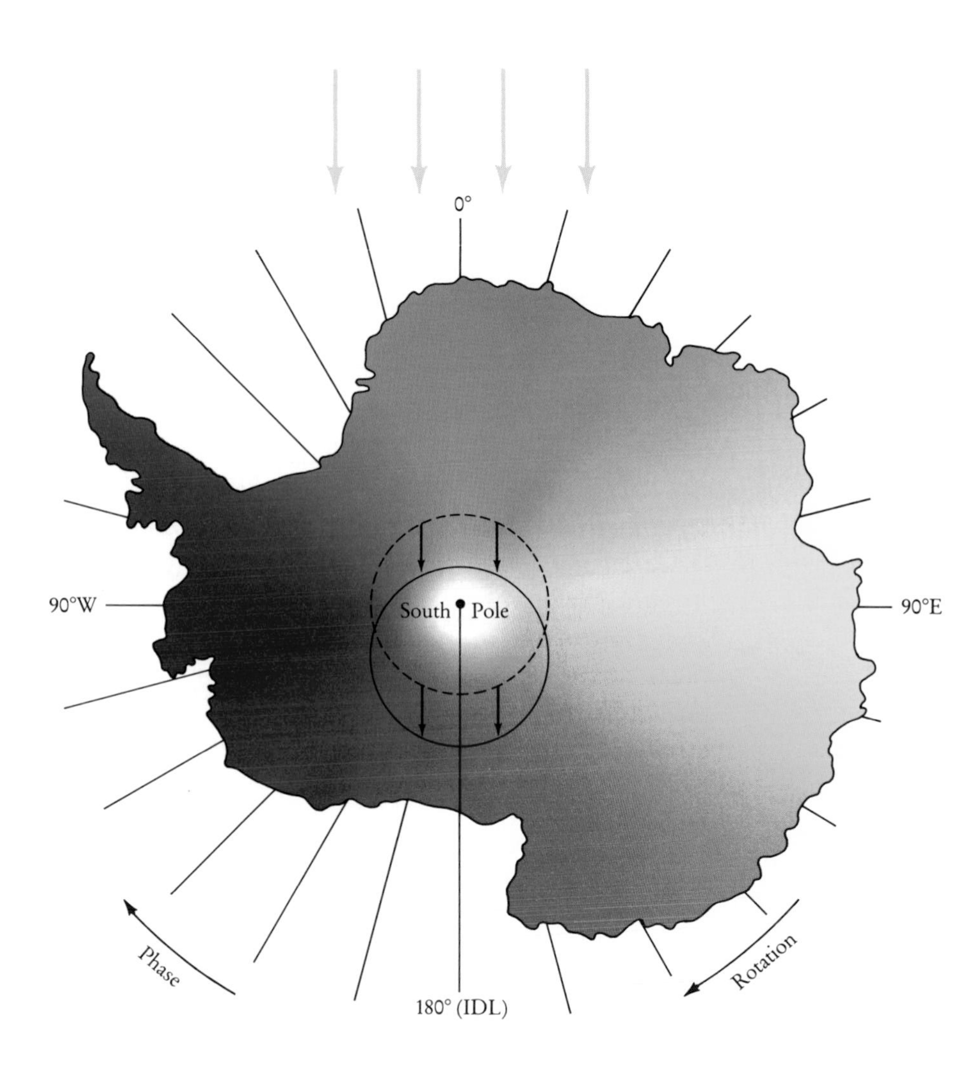

New phase
Old phase

On the Antarctic Continent, time zones converge to the pole. The colored zones are stationary while the continent rotates through them. A circle of clods of earth (dashed ring) samples every time zone. Physically displacing all of them away from Greenwich by some distance (new positions shown along the solid black ring), each clod's circular orbit around the pole is altered both in amplitude and in phase. The new phase reached (the new time zone or color) depends on the original phase as graphed to the right of each map: odd resetting for small displacement, even resetting for larger displacement.

ess. From the perspective of a stationary observer in space—one of the equatorial geostationary weather satellites, for example—each clod exhibits 24-hour rhythms of distance and apparent size as it rotates along the dashed cycle.

Now imagine that some great disturbance suddenly pushes every clod in the same direction, as though signals from the Greenwich Observatory were to suddenly acquire force and shove each clod along parallel lines across the icy surface. At the end of this push each clod has a new rhythm of distance from the observer; it still repeats every 24 hours, but

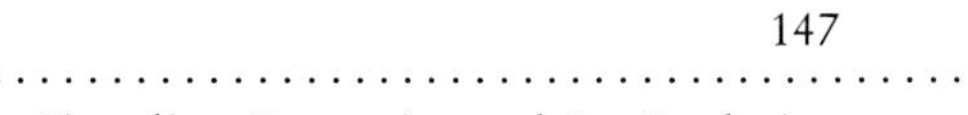

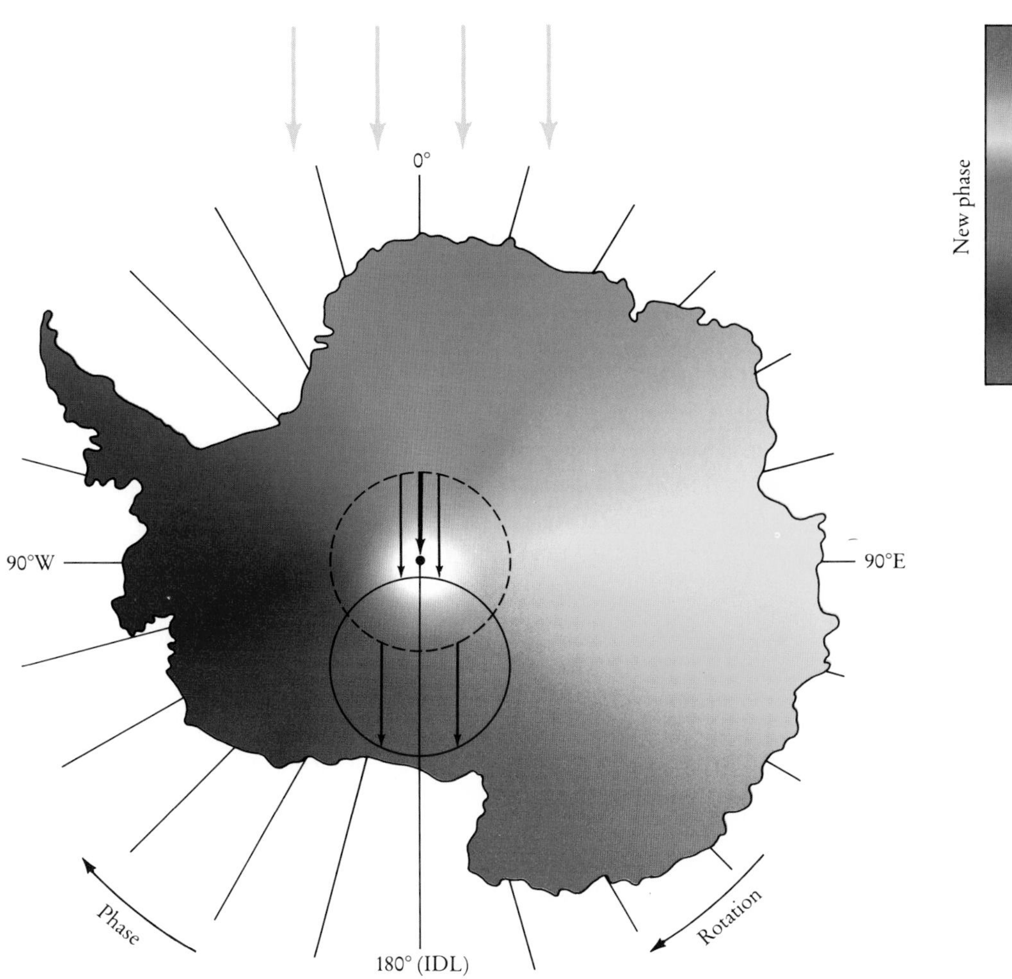

the hour of closest approach has changed. The phase of the rhythm has been reset. (Incidentally, notice that the range between minimum and maximum distance—the amplitude of the rhythm—has also been reset.) From each initial old phase, a clod was instantly reset to the corresponding new phase. What is that new phase, and how does it depend on the old phase?

The old phase can be read directly from the time zone line (or color) on which the clod was initially situated, and the new phase can be read from the time zone line (or color) on which it ended. Graphing new

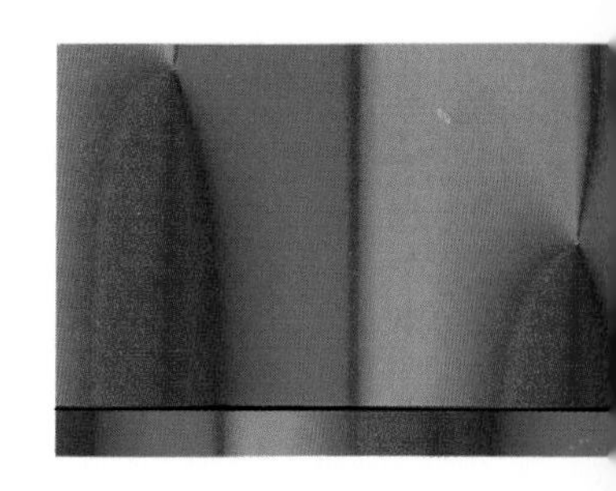

phase against old phase produces a resetting curve. If the displacement was slight, this curve is almost a straight diagonal line: new phase = old phase. Larger displacement bends the curve into a sinuous wiggle about the diagonal, but leaves it still of odd type. It remains odd type as long as the circle of clods still encloses the South Pole, so that, in going once around the new circle, the new phase values also run once forward through all time zones.

Consider now how things change if the stimulus was sufficient to push clods all the way across the pole. Now a clockwise trip around the new ring of clods leads through only some time zones, forsaking others altogether. Each zone encountered is encountered twice, once with new phase increasing as before, but then once again with new phase decreasing. This is quite different. After this larger stimulus, the new phase depends smoothly on old phase in the fashion we called even resetting.

If one thinks in terms of phase *shifts,* the smooth even resetting pattern could be mistaken to disclose a discontinuity in the mechanism of resetting. Neighboring clocks straddling longitude 0° are transported across the pole to straddle the international date line. They had practically the same old phase; they now have practically the same new phase; and they passed through practically the same places en route. But viewed in terms of phase shifts, one advanced through half a cycle and the other delayed through half a cycle. This verbal distinction might make sense if they had to advance or delay along quite different routes, for example, if they literally traversed the original dashed cycle clockwise and counterclockwise, respectively, to end up together again on the far side. But in terms of the neighboring clods, not even that much distinction can be made between a big advance and a big delay. They are almost indistinguishable parts of a smooth pattern of resetting that differs from the familiar odd pattern, not by a discontinuity, but in a more subtle way. Even resetting differs from odd by a topological characteristic that is obscured in translating the raw data from observed timing of a rhythm (old phase, new phase) to an inferred change of timing (new phase minus old phase plus or minus a cycle = phase shift).

Where does the singularity fit into this picture? There is only one point on the southern hemisphere where the time zone is ambiguous, where a clod would be in all time zones at once, or in none of them: at the pole. At the pole, the rhythm's amplitude is zero—the clod does not change its distance from our observer. How does a clod end up at the pole? Only by starting in a particular time zone (the Greenwich meridian, the vulnerable phase) and receiving a push of critical length. This is the singular stimulus of critical size that divides odd resetting from even, and which annihilates the rhythm if and only if it is applied at the unique vulnerable phase.

This parable could be told as well about corks floating in the ocean, in a ring around any one of the many amphidromic points: the range and timing of their tidal rhythms is identically affected by a wind that blows them all far enough. But any such quaint analogy between geophysical arrangements and the timing of biological events recalls a warning issued by Aurelius Augustinus, the Christian bishop of North Africa fifteen centuries ago. His counsel concerned astrologers, the "mathematicians" of his day who sought connections between astronomical cycles and the processes of life:

> Beware of mathematicians and all those who make empty prophecies. The danger already exists that the mathematicians have made a covenant with the Devil to darken the spirit and to confine Man in the bonds of Hell.

Let us then see if these particular prophecies must remain empty. We will first see how well data fit into them, and then see whether they have some meaningful interpretation.

A Quantitative Description of the Time Crystal

The geometry of the polar metaphor is so simple that new phase can be written as a simple mathematical function of old phase and stimulus size. The resetting surfaces in Chapter 4 were in fact drawn by a computer using nothing more than that equation. With only a little gentle tailoring, so were the illustrations in Chapter 5 that represent resetting in the fruit fly cued by a light pulse. Those 1574 resetting measurements happen to conform to the calculated values to within 2 hours (8 percent), close to the ± 1.5-hour reproducibility of any phase measurement in that experimental system. Roughly the same is true of the drug-cued resetting measured in *Gonyaulax,* in Chapter 6.

Though not an empty prophecy, in that it does quantitatively describe the pinwheel experiment, this metaphor still does not *explain* the circadian clock's style of resetting on cue. For the moment it serves only as a way to visualize the pattern and its singularity in terms of something more familiar. However, the same pattern, even exactly the same diagrams, will recur shortly, reinterpreted as a description of the clock's inner dynamics.

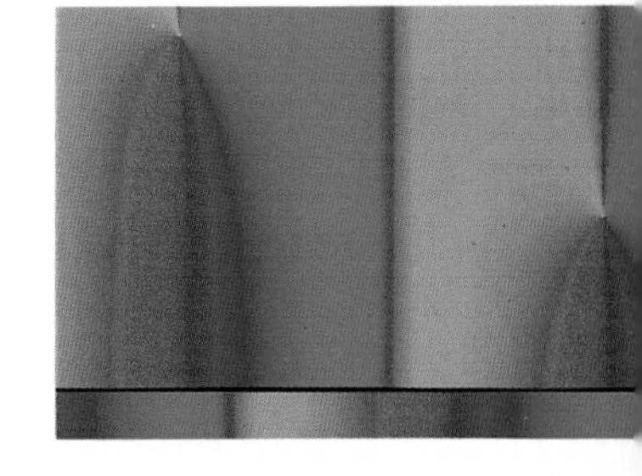

The Shapes of Chemical Dynamics

About twenty years ago, the format of description long favored by engineers concerned with stability and oscillation in complex systems helped biologists formulate a new vision of rhythmic dynamics. It was so much more realistic than the vague and arbitrary "models" then in vogue that many an old debate was simply forgotten and experiments took a new direction. It was realized that topological methods would allow such analysis to proceed even before the clock's mechanism is deciphered—as long as inquiry was restricted to features of clock behavior that are fairly independent of a particular mechanism. It turns out that the circadian clock's most essential behavior, its ability to reset on cue, is one of these.

The basis of this vision, familiar to physiologists since at least 1960, is that the circadian oscillation must derive ultimately from some kind of continuous interplay among two or more distinct changing quantities: membrane potentials, ion fluxes across membranes, or concentrations of enzymes or of their substrates, to name but a few plausible candidates. In oscillating systems, that interplay forces these quantities all to bobble up and down in a characteristic sequence around their average levels. Biophysicists studying cardiac pacemakers, biochemists studying rhythmic sugar metabolism, and physical chemists studying periodic reactions are accustomed to write differential equations to describe how each quantity's rate of change depends, moment by moment, on the values of all the quantities involved.

A graph of one such quantity against any other shows both quantities changing in time along a path called a *trajectory.* The trajectory might cross over itself (as in the graph on page 122) or it might not. But whatever happens, when the system has settled into a regular rhythm, the trajectory closes in a loop.

What would you see if you could plot all the essential quantities (not just two) against each other simultaneously? If a dozen or more distinct quantities are involved, as in the known mechanisms of sugar metabolism and other well-studied chemical oscillators, then you need graph paper with a dozen or more independent plotting axes, an exotic item available only inside computers and difficult to visualize. But the trace, being periodic, is still simple: it is some kind of closed loop, or cycle. In this complete representation, the system's path never crosses itself. Why not? Because each point on this multi-dimensional graph paper represents a unique combination of all the relevant quantities; by definition, they will force each other to change in a unique way as described

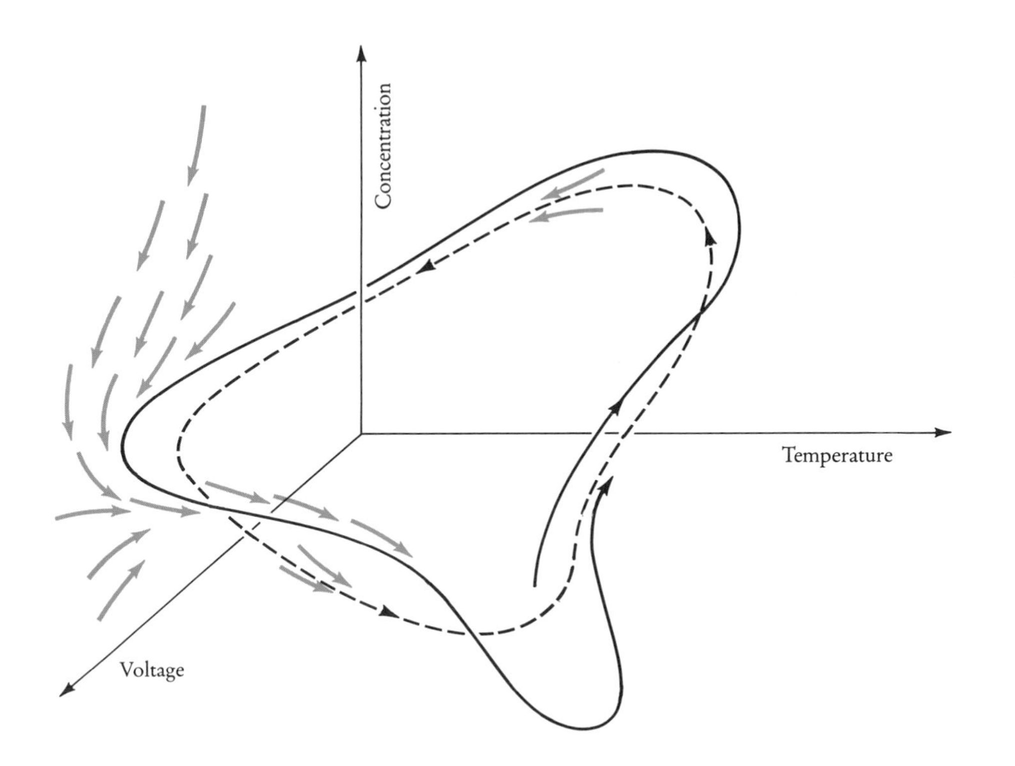

Here, a changing system is described schematically by its temperature, voltage, and some chemical concentration. If the state of the system is represented as a point in this coordinate space, its dynamics would be represented by the path of that point. The path is steered by a direction field: at each state (point) all three quantities have unique rates of change, which define a tiny arrow to a neighboring state. If the system is periodic, its path closes in a loop (dashed line). The direction field at any point off the loop eventually moves that point back onto the loop.

by the pertinent differential equations. That change in the next instant is a small displacement of all the quantities: a tiny arrow uniquely defining the path followed by the system's representative point.

The field of little arrows, one at every point, collectively constitute a wind that blows the system along. We know that the system has a natural tendency to repeat its standard sequence of changes, so we know that the wind blows along a loop. Moreover, the wind blows *onto* that standard loop, so after almost any small disturbance, the system reverts toward the loop, getting closer and closer as time increases. This loop is called an *attracting limit cycle* because, strictly speaking, it is reached only in the limit as time increases to infinity.

A changing system's full behavioral repertoire can be comprehended in these terms as a single timeless snapshot of its "temporal morphology." This is only a geometrical way of saying three things: that each quantity's rate of change (the wind) depends only on the instantaneous values of all pertinent quantities; that the wind at neighboring states is similar; and that the system can be regarded as a single homogeneous

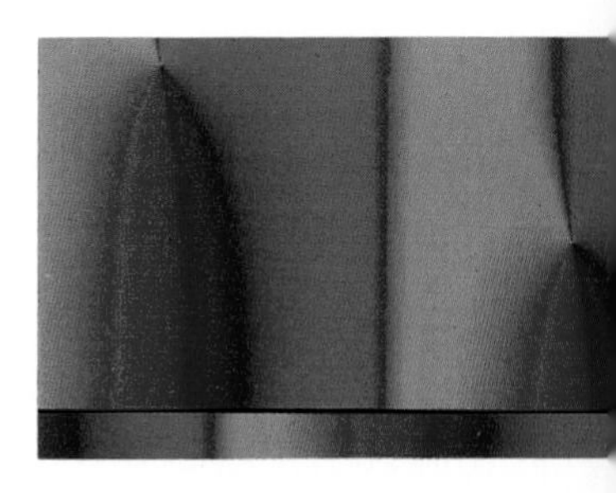

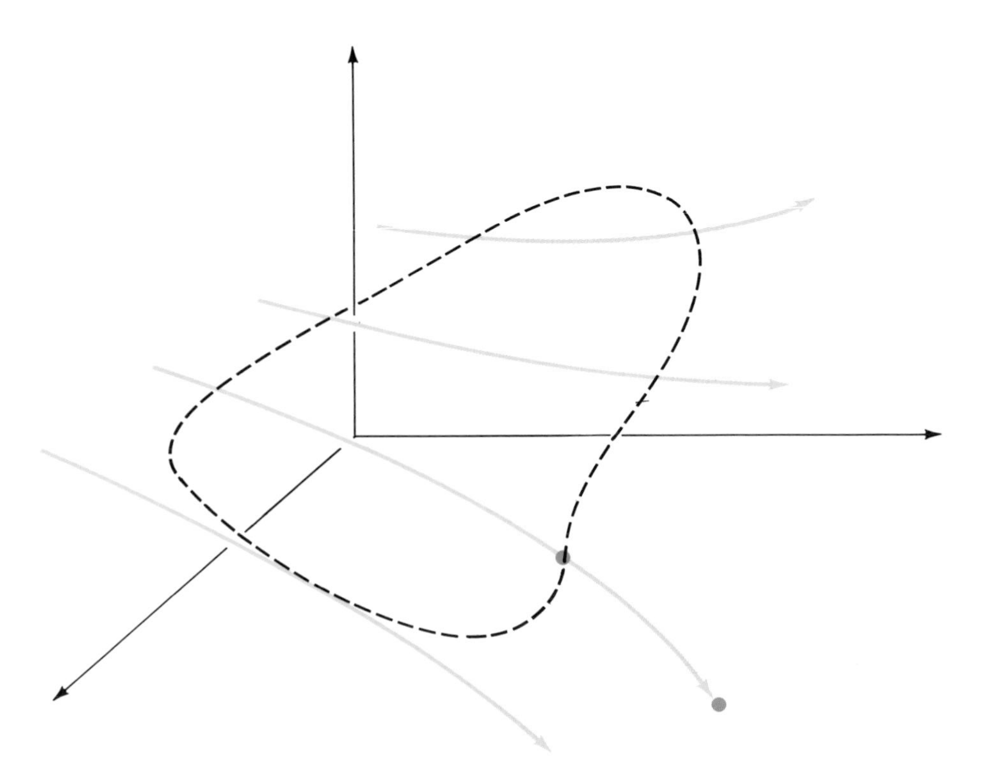

Under changed conditions any system generally behaves somewhat differently, so the arrows of the direction field are arranged differently. Here a change in conditions—a phase-altering stimulus—is represented as a "wind" that sweeps across the closed loop, driving the system-point off the loop.

unit (a point). A uniformly stirred chemical reaction, for example, satisfies these requirements; if unstirred and capable of behaving differently in different regions, it might not. Or if reaction rates change substantially in response to arbitrarily small changes of concentration, it definitely does not. Other than that, there is almost no "theory" involved here: this is just a way to represent observed behavior in visual format so that subtle implications become accessible to one of our most powerfully developed intellectual tools: visual intuition.

In these terms, a stimulus is a momentary alteration of the differential equations, a change in the anatomy of the wind. To a system stably cycling, the stimulus is a shift of the wind, nudging a cycling point off the attracting cycle out into the surrounding space. From wherever the system finds itself when the changed wind returns to normal (when the stimulus ends), it follows the normal wind back toward the cycle. The unperturbed system (a control experiment) may be imagined to have continued cycling like a sailboat going round and round a race course,

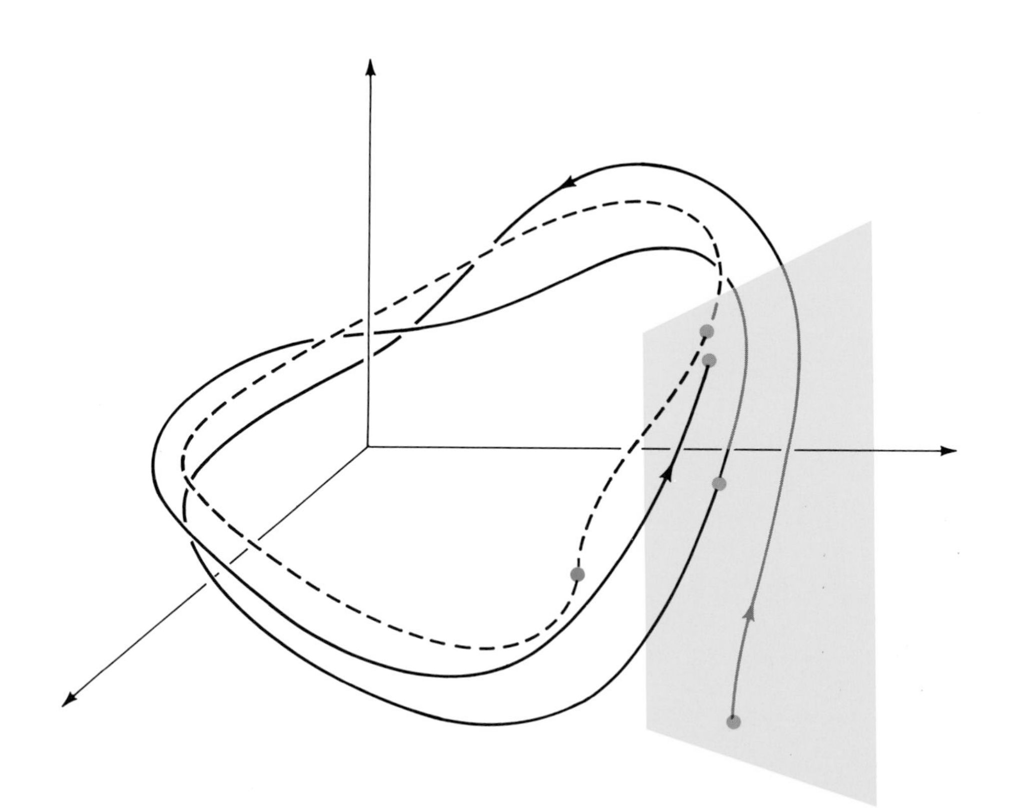

After having been blown off course by the stimulus-wind in the preceding illustration, the system again finds itself under the influence of the old direction field. At intervals of one cycle period it appears at the red points, eventually converging to the attracting cycle (dashed loop) at some new phase. A replica of the system before perturbation is still rounding the loop and reappearing at the same old phase each time (blue). The yellow slice through this space contains all states from which the system would end up at the red point on the loop after a whole number of strobe flashes. There is one such isochron sheet for each point on the loop.

while a perturbed replica exposed to the stimulus was blown off track, and begins to return when the wind reverts to normal. Generally, it re-enters the course somewhat ahead or behind the control.

We now have an interpretation for the role of the wind "stimulus" in the Antarctic metaphor. What about the time zones? Suppose a flashing light permits us to look only at times that are multiples of one cycle after the stimulus ends. In each flash of the strobe we find an undisturbed system back at its original old phase, and a reset replica getting closer and closer to the track at some new phase. The new, reset phase depends on where the system was when the wind first changed, and how strong and persistent the change was—in other words, on the timing and size of the stimulus. The details of this dependence depend on particulars of the model, namely, the geometry of the cycle and the direction of the altered wind; but its overall pattern is largely independent of such particulars. This independence derives mostly from a topological theorem about the ways in which a state space can be sliced up into sheets (of one

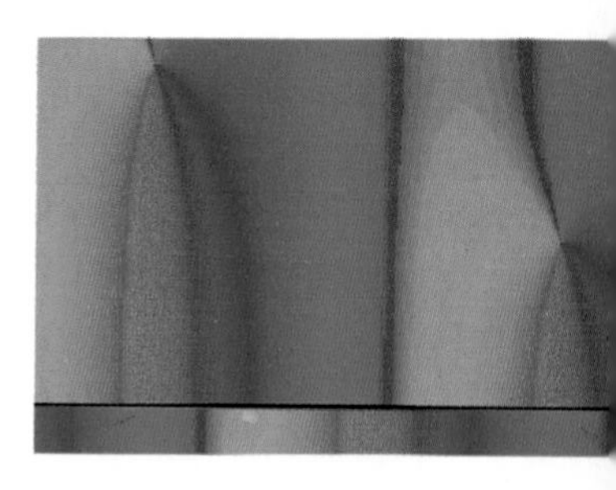

less dimension than the whole space) with a peculiar property. The yellow surface in the illustration on the preceding page is one such sheet: a system-point found anywhere in that sheet during a strobe flash will have cycled back to the sheet in time for the next flash, and eventually will be seen again and again at the red point on the sheet where it is punctured by the attracting cycle. Such sheets are called *isochrons,* like the isochrons of new phase on a rectangular graph of old phase and stimulus strength. The isochrons mark time zones in state space, and just like the 24 conventional time zones, they necessarily come together somewhere inside any cycle that penetrates them all in orderly succession. That somewhere must be phaseless.

The Trajectories of a Pendulum

Consider a simple pendulum in these terms. Actually, the simplest pendulum does not tend back to a unique cycle, since it has no preferred amplitude; in fact, it will eventually run down and hang still. But imagine a pendulum like the one hanging beneath a cuckoo clock, attached to some hidden mechanism that forces it gently toward a standard range of swing. At any instant, the state of the pendulum can be expressed without ambiguity by only two quantities: the instantaneous velocity and position of its end point. Both change rhythmically: at the far right

Velocity and position change rhythmically as a pendulum swings. Plotting one against the other reveals a circular trajectory like those considered more abstractly in the preceding illustrations.

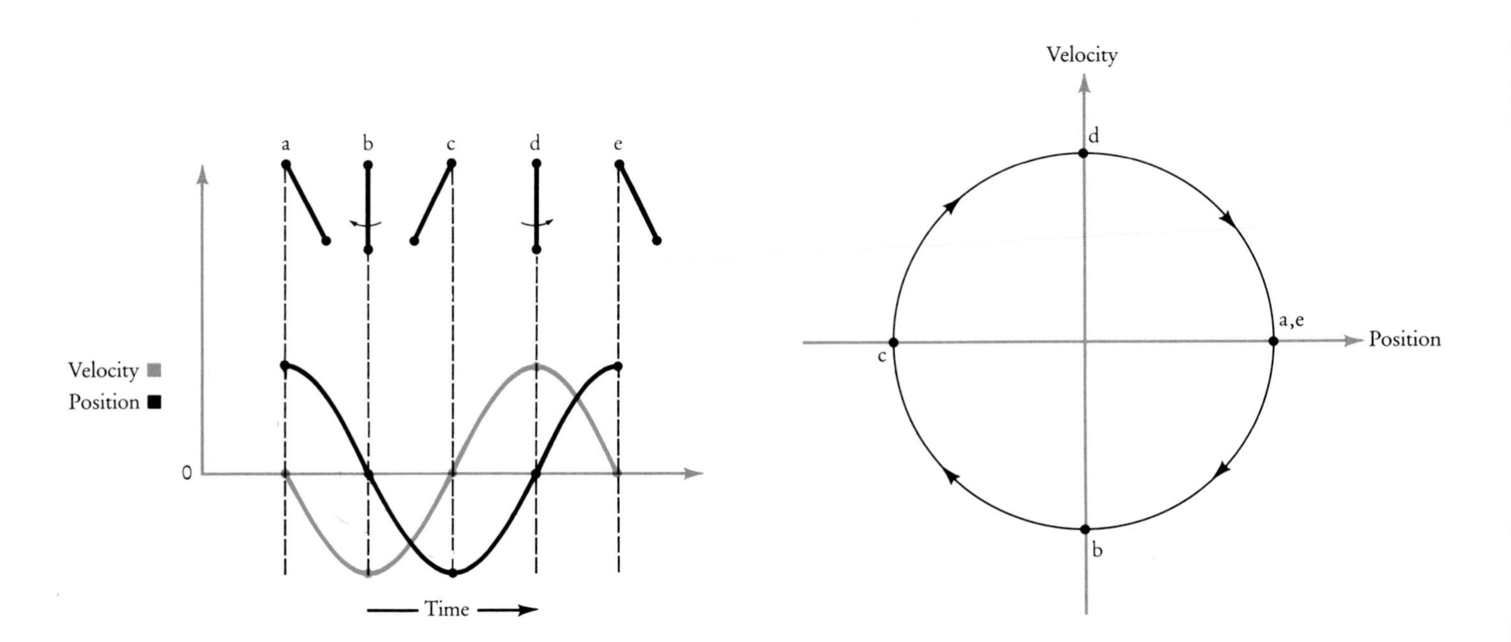

extreme, velocity is zero as the bob is about to fall back to the left; at the bottom of the arc (middle position), velocity is as fast leftward (negative) as it will ever be; at the extreme left, velocity has again fallen to zero; back at bottom, velocity is maximum rightward (positive); back at the extreme right, one cycle is complete. The path traced on a position-velocity graph is a circle. If the pendulum began by swinging with greater vigor, the extremes of position and velocity would be greater: it would traverse a bigger loop in the same period. As its amplitude regulates toward standard, its loop shrinks toward the standard circle.

What if you gently tap the pendulum with your hand, bleeding off a little of its speed? On the diagram below, an abrupt change of speed is represented by a vertical displacement; if the bob's position is not also

Instantly reducing the speed of the pendulum changes its state from A to B. From B it follows a spiral trajectory back toward the attracting cycle. At unit intervals, it reappears at points along a radial isochron (yellow) as it approaches C. Other isochrons are also radial. They all come together at the equilibrium state: neutral position, velocity zero, oscillation amplitude zero, phase ambiguous.

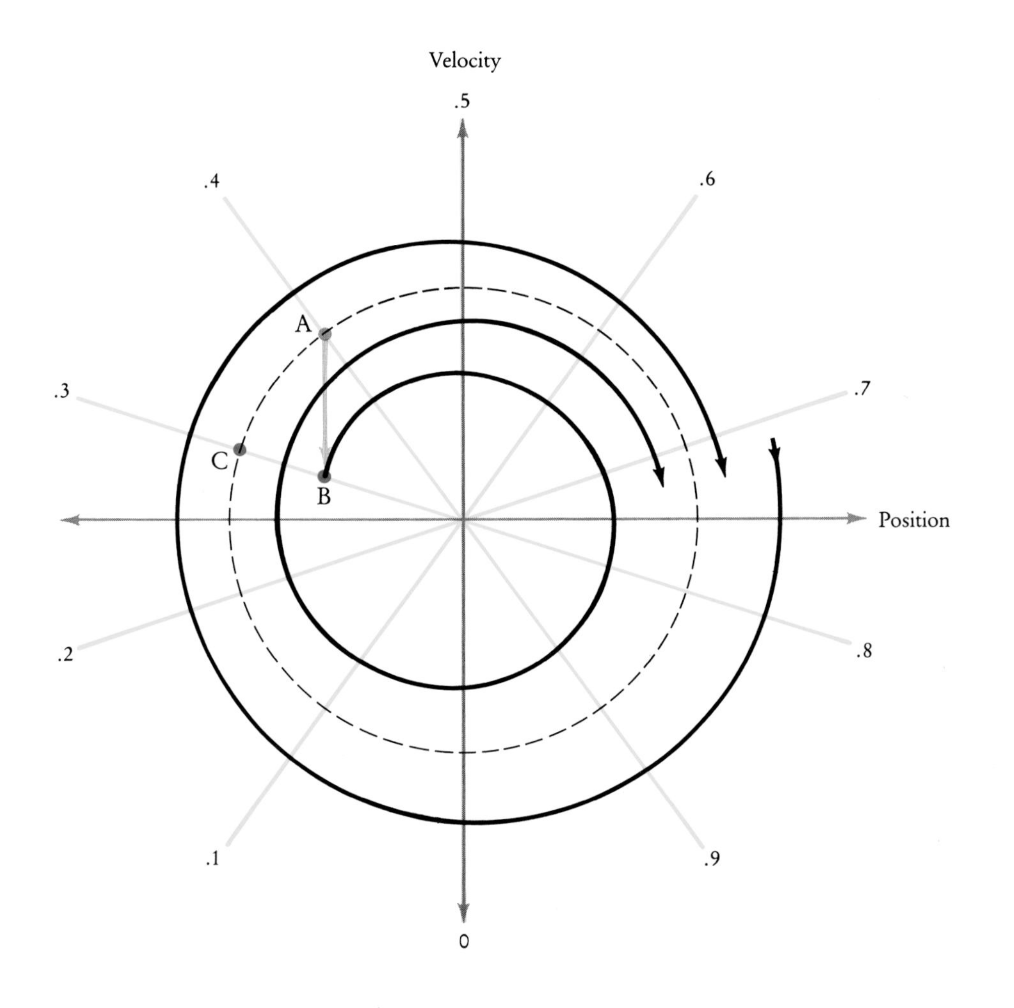

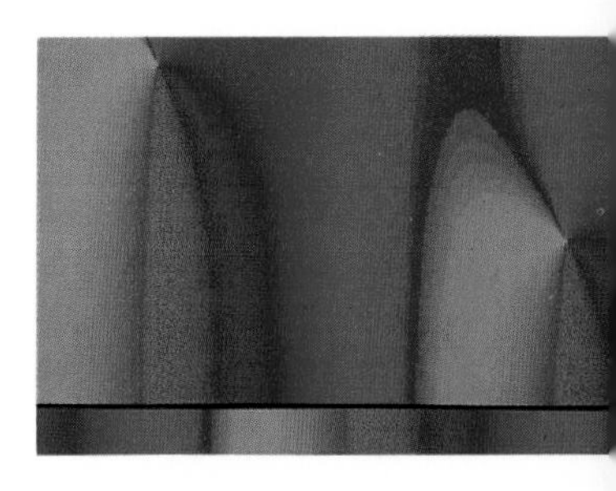

changed instantly, decrease of speed is simply a downward displacement from A to B. From B, the pendulum follows the "wind" (direction field) through B—along a smaller but growing loop.

Notice that the clock starting from B will reach the top of its cycle (middle position, maximum speed to the right) somewhat later than it would have had it continued unperturbed from A. It has been delayed. By how much? We find out by watching as a strobe (set to the pendulum's period) flashes until the clock settles to a fixed point on the attracting cycle. From flash to flash, it seems to hop along a radial path, because in each interval between flashes it goes exactly once around, only changing amplitude a little. Started from anywhere along that radial line, the clock would converge to the same phase on the attracting cycle. This line is an isochron in the state space. The other isochrons are other radial lines, each one crossing the attracting cycle at its own unique phase. Let the bottom position on the cycle be called phase zero. The isochron through that point is called the zero isochron. The rest are numbered in order around the cycle.

Finally, we can read the new phase associated with any bump given to the pendulum. Let the pendulum be initially at A. Its old phase is read from the isochron through A: 0.4. Perturb it at that moment with whatever kind of stimulus, in this case a decrease of velocity to B. Read the new phase directly from the isochron through B: 0.3.

If this experiment is repeated many times with the same stimulus, given at every possible old phase A1, A2, and so on, then the corresponding new phases B1, B2, and so on are arranged along the solid black circle in the diagrams on the facing page. In the situation on the left, notice that every isochron is crossed. Just as in the Antarctic metaphor new phase varies through one cycle as old phase varies through one cycle; we have odd resetting. On the right, a bigger stimulus is represented. Here, each new phase is encountered from two different old phases; this is even resetting. All clocks are confined now within a wedge of isochrons, with more of them near the edges than near the middle. In contrast, had they initially been not uniformly spread around the cycle, but near equilibrium, the stimulus would have left them all near the midrange isochron. This is the basis of the experimental test to distinguish between the two interpretations on page 102.

Notice in this pendulum analogy that the isochrons converge radially to a point inside the cycle. If the dynamics is more complicated, then the isochrons needn't be exactly radial and their convergence needn't be a point; and if the dynamics involves more than two independent quan-

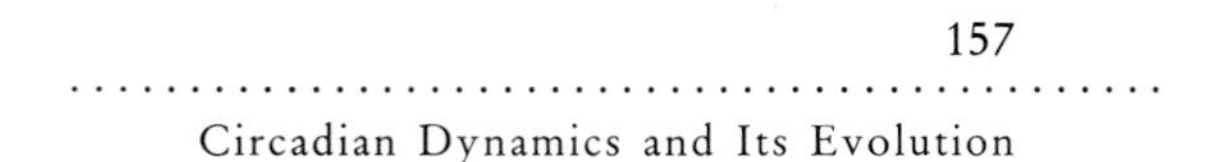

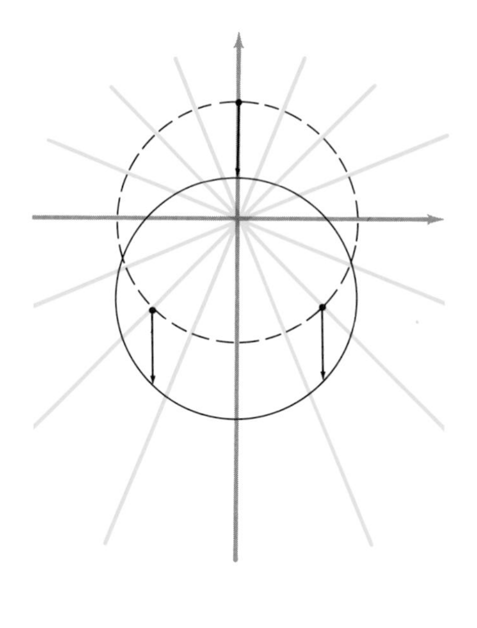

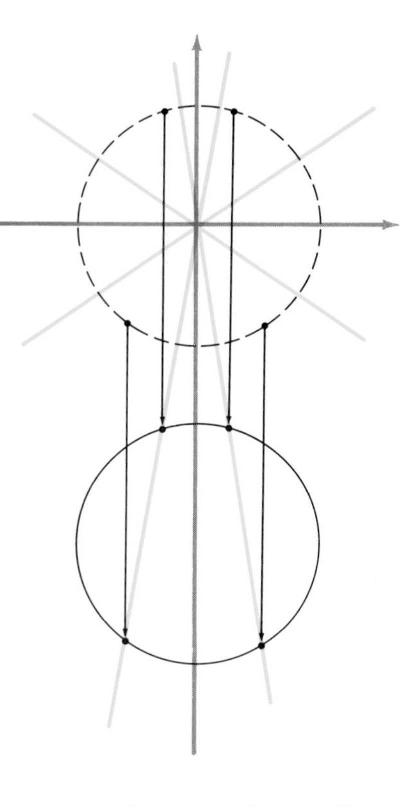

Left: Starting from any phase on the attracting cycle, an impulse of fixed strength changes the velocity downward by a fixed amount. The new states lie along a new ring enclosing the convergence of isochrons much as in the Antarctic metaphor; as old phase increases through one cycle, the new phase also executes one full cycle. This is odd resetting. *Middle*: A stronger impulse moves the state farther. As old phase is increased, a new state changes along a new ring that no longer encircles the singularity, so the new phase increases and decreases through a limited wedge of isochrons. This is even resetting. Each isochron (new phase) is reached from *two* old phases. *Right*: The middle panel is dissected to show 20 individual clocks initially equipaced around the cycle. After even resetting they all lie within a three-hour arc of new phases. Most lie near the extremes of the arc, with fewer spanning the middle hours. In contrast, clocks initially near an equilibrium point (blue) would all be perturbed to the same new phase, in the middle of that arc.

tities (dimensions), then the convergence needn't be only inside. But it can be shown mathematically that the isochrons must converge, and that odd resetting changes to even when the stimulus becomes big enough to push the closed loop all the way across that locus of convergence. If that locus is a mere point, then a stimulus of that critical size, applied at a particular old phase, the vulnerable phase, leaves the system at the convergence point. What happens then? There are many possibilities for different clock mechanisms, but they all have in common that the wind through that region does not blow back onto the attracting cycle, and that by following winds nearby, the system could hop (under

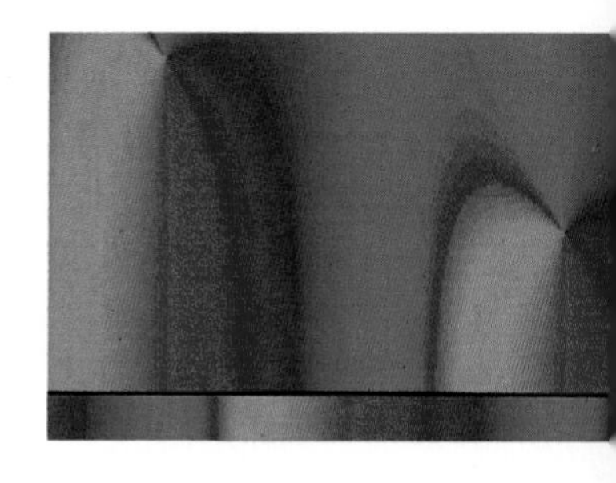

strobe illumination) to any part of the cycle. So the outcome must be arrhythmia or an unpredictably delayed recovery to unpredictable new phase. If the convergence is not a mere point, it is nonetheless commonly a locus of two less dimensions than the whole state space, and everything works out the same. It also sometimes happens that the convergence locus is bigger, having only one less dimension than the state space (for example, the border of a phaseless region in a two-dimensional state space). In this case, the new loop encounters the convergence locus over a range of stimulus sizes, and resetting curves in that range are discontinuous. But below that range is odd resetting, above it is even resetting, and along its fringes all phases coexist.[1]

Models and Mechanisms

In the case of a simple pendulum, the singular point represents an equilibrium state. Does this suggest that circadian clocks have an equilibrium state too, perhaps a biochemical equilibrium? Perhaps, but not necessarily. If circadian clocks involve the interplay of more than just two quantities, then the state space is at least three-dimensional, the isochrons are at least two-dimensional, and their convergence is at least one-dimensional: not a unique state, but a trajectory or a bundle of trajectories distinguished only because they do not lead back to the cycle that attracts nearby trajectories. The phase singularity in such cases is manifest as an arrhythmia, but not necessarily as quiescence.

These abstract diagrams are the conceptual substance of the polar metaphor that opened this chapter, and they have been used successfully to analyze many kinds of cyclic systems. In dealing with a chemical clock, we would use not position and velocity coordinates but the concentrations of reactants. In electrophysiological pacemakers, the coordinates are a voltage difference across the cell membrane and the fractions of several kinds of membrane ion channels that are open at the moment. Details have been worked out experimentally with great thoroughness in a dozen systems. But for circadian clocks, the picture still seems so abstract that it may seem little more than another puzzling metaphor. What kinds of mechanism can be described in these terms, and what kinds cannot? From the diversity of clocks now known to reset on cue in the manner first predicted from this diagram, it seems clear that it is vaguely compatible with an abundance of different mechanisms. Particular mechanisms lead to particular isochron shapes, particular critical stimulus modality and size, and particular timing of the vulnerable phase. They accordingly lead to phase singularities in the graph of old phase and stimulus size, which consist of one or many isolated clockwise or counterclockwise points, or one-dimensional arcs, or even boundaries of two-dimensional black holes. The arrhythmia following a singular stimulus may consist of mere quiescence, of randomly timed return to the same cycle, or of more complex behavior. Such distinctions about the resetting behavior may prove useful to distinguish between hypothetical clock mechanisms, but not before the field matures enough experimentally to provide such realistic and quantifiable alternatives.

Lacking even that much guidance from the laboratory, we can still ask at least whether any classes of mechanism are hopelessly *in*compatible with the style of resetting that seems characteristic of circadian clocks. Yes, some are, but they are only very literally interpreted simple clock models that involve a single quantity (as in "A Parable of Three Clocks," in Chapter 3), or that involve fundamental discontinuities (as in "Nuclear Fission in Slime Without Cells," in Chapter 7). In some essential respect, "clock" is an inappropriate metaphor; familiar clocks have no vulnerable phase at which they can be reset randomly to any phase whatever by a stimulus of critical size. And familiar clocks have nothing like an amplitude; they have only a single dimension of the internal state space and accordingly must always be at *some* phase in the standard cycle. Although one-dimensional and even discontinuous models dominated (and often misled) the practical design and interpretation of experiments for decades, probably no one is surprised to discover that their strict interpretation is fundamentally inappropriate.

Yet even such clock models might become acceptable if the biological rhythms observed turn out not to really represent the unit mechanism but rather the collective behavior of an inhomogeneous "clock shop." From this point of view, it seems thoroughly unsurprising that the resetting behavior of fruit-fly circadian clocks has since turned out to represent a principle of wide application, typical of diverse biological clocks and compatible with many attracting-cycle mechanisms.

All the examples in this book can be presented in terms of one or another hypothetical model of this class. The trouble with this vision, though, is that it is altogether *too* easy to construct attracting-cycle models. Such modeling rapidly came into vogue once the idea was broached among circadian physiologists around 1960.[2] The coordinates have different labels in different schemes, and the phase singularity appears in many distinctive guises, but experimental work is not yet refined enough to distinguish between them. At present, the interacting

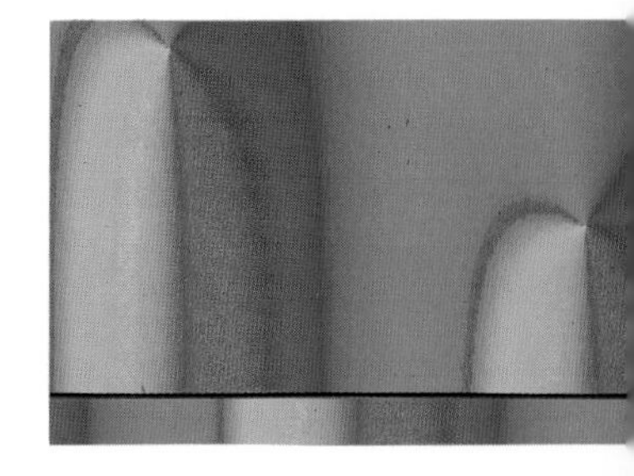

hypothetical quantities, their high-dimensional coordinate space, and its attracting cycle all remain gratuitous constructs, since not one such quantity can presently be named or measured in any particular circadian clock. In contrast, they are all known and quantified in several kinds of chemical oscillator, including sugar metabolism, and in several kinds of nervelike pacemaker. Thus, the situation for circadian clocks may be expected to change, but it takes time.

Why Do We Have Clocks?

Who knows for certain? Who shall declare it? Whence was it born? Whence came creation? The gods are later than this world's formation; who then can know the origins of the world? None knows whence creation arose and whether he has or has not made it; He who surveys it from lofty skies, only He knows . . . or perhaps He knows not.

RIG VEDA X:129

Scientists have proposed many different accounts of the evolution of biological clocks. But it is impossible to be sure about things that happened in the first living organisms. We don't know what those organisms looked like, nor the exact chemistry of their world. But we do know with certainty that the sun rose and set regularly, periodically assaulting the seas with energetic photons that living organisms eventually learned to tame for photosynthesis. And we know this result: that living things on earth today mark time innately by units of 24 hours, give or take an hour. They don't need to learn this rhythm, they don't need cues, and they don't forget. It's built in.

Though we cannot account for this rhythm, neither can we silently bypass such a fundamental mystery. Abandoning any pretense of certainty, I relate a few *Just So Stories* about how natural selection may have favored those individuals whose internal dynamics matched their home planet's daily rotation period better than their less successful and now extinct competitors.

The first story turns on a timing system designed to overcome the disruptive effects of sunlight on the delicate machinery of the cell. The second story focuses on an early cell's daily cycle of activities, imposed at first only by the daily cycle of light and dark, warmth and cold. A third starts with an imaginary primitive cell metabolizing steadily in a constant environment, and asks how stable that steadiness could be, even without the impact of sunrise and sunset.

First Dream: Dawn Warning

Suppose it were desirable for the earliest cell to coordinate its biochemical activities on a daily schedule. In a stimulating speculative essay, the evolutionary biologist Colin Pittendrigh suggested advantages for cells that could anticipate the rising of the sun:

> Little attention seems to have been given . . . to the problem of visible radiation as an energy source that threatens organization. . . . The majority of the cells' constituents are colorless; and uncontrolled activation by the visible is thus excluded. It may well be that in the history of the cell there has been selection for colorless molecules, but . . . for some functions, colorless molecular devices have not been found. . . . No attention seems to have been given to the consequences of illuminating these molecules whose color is without obvious function. . . . Some subroutines in these cells' overall tasks are impaired by the activation of molecular piece-parts in the flood of visible radiation to which it is subjected each day. To that extent, the routine delegation of some chemical activity to the recurrent darkness of each night would be an obvious escape from the photochemical threat to organization.[3]

If a timer is needed to anticipate the dawn, it seems reasonable to expect that the simplest timer would arise first. As in the slime mold (see Chapter 7), some chemical byproduct of metabolism might accumulate during the dark until a threshold is reached. To have a clock, the cell need only use that threshold crossing to switch from metabolism appropriate in darkness to metabolism appropriate in the glare of sunlight. If the product dissipates during daylight, then we have an hourglass and the means to turn it over. (Equivalently, a product might decay in the night and be restored to a standard concentration by the end of day.) A sophisticated accumulator might even protect the rate of accumulation from day-to-day vicissitudes of temperature and light intensity. A still fancier mutant could adjust the threshold if it is reached a little too early or too late today; then the timing would be more nearly on target tomorrow.

Such a biochemical interval timer would be analogous to an hourglass started up by dawn or dusk. Yet the biggest mystery about circadian rhythms is their spontaneity. Why do biological clocks still work when

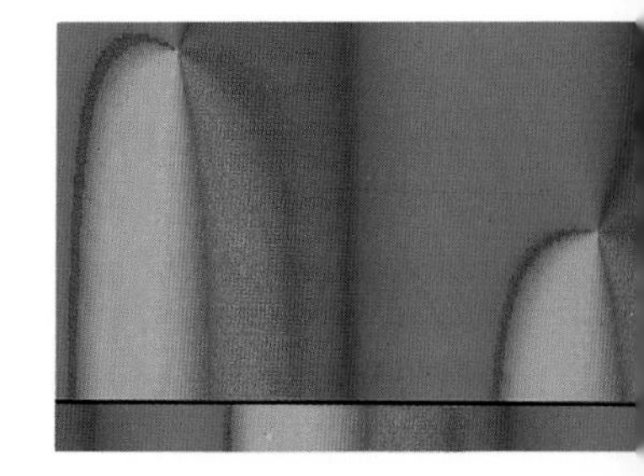

hidden from the sun? As Sherlock Holmes would say, "Depend on it: there is nothing so unnatural as the commonplace." The circadian timer usually recycles smoothly at 24-hour intervals despite the lack of obvious selective pressure for such a contrivance. A primordial hourglass easily could have mutated to turn itself over, but given the regularity of sunrise and sunset, such cells would have enjoyed no particular advantage over nonspontaneous competitors. Except in Dracula and other nocturnal mammals that sally forth only by night from perpetually dark lairs, circadian clocks seldom get a chance to show whether they are spontaneous. So who needs spontaneity? Perhaps it is not an adaptation at all, but an accident of neutral selective value. If so, it hints that the origin and mechanism of the clock may lie in some further considerations.

Second Dream: A Ring of Habit

Suppose that clock evolution started not with the need for an interval timer, but with a pre-existing cycle of biochemical activities, driven by the daily alternation of light and dark, of warmth and cold. With many metabolic pools driven into daily ups and downs, the cell could be regarded as performing a daily round of chores: photosynthesis by day, heat prostration by evening, perhaps reproduction (cell division) by night, starvation by morning.

On a rotating planet, environmental rhythms intrude forcibly into the cells' internal affairs, forcing internal rhythms. A cell needs no additional clock. But if it could anticipate these regular changes, it might improve its own efficiency by optimizing the transfer from one metabolic preoccupation to the next. To avoid waste, to anticipate changes and bridge them smoothly, to resist diversion during an eclipse or the passage of a cold front, the successful cell would accumulate modifier genes that ease each hand-off of control during this imposed cycle. The descendant of such a cell might be surprised to discover, two billion years later when some scientist first puts it into a constant laboratory environment, that it still shuffles its way spontaneously through almost the same cycle. It might seem hard to exclude from a laboratory the incessant geophysical rhythms that pervade almost every aspect of life on a rotating planet, but it turns out that shielding just from changes of light and temperature enables organisms to free-run at their own individual periods, typically several percent different from 24 hours. Yet the ability to do so has never been revealed before and so could never have been selected for!

Imagine that, as in Fred Hoyle's best-seller *The Black Cloud,* the earth is threatened with perpetual night. The planet will still rotate, but sunrise and sunset will no longer cue the daily rounds of work and play. Will city streets still be choked with rush-hour traffic twice in every 24 hours? Probably so, because cities are full of little 24-hour oscillators called wristwatches that serve many of the same purposes as sunrise and sunset for city dwellers.

But suppose for a moment that watches were not made to start over at 0 after 12 or 24 hours: suppose they just kept linear time since the turn of the century. Cities would then go uncued by any 24-hour timer. Would the ancient alternation of business and night-life then decay into a steady hum of traffic, sleep, business, lovemaking, and crime unstructured by the throb of diurnal rhythms?

Or might Hoyle's global nightfall possibly reveal that the 24-hour period has somehow become internalized, that it maintains itself without cueing in cities much as it does in individual organisms? In many ways a city is like a single organism composed of many self-reliant but interdependent cells (persons), organized into tissues (businesses, agencies) that fulfill one another's needs. The newspaper can't release its product continuously; it gathers stories, prints a huge pile of copies, and only then is it efficient for the delivery trucks to back up to the loading dock. Release of political, financial, and business news starts the day in downtown offices as reporters arise from late sleep to harvest today's new surprises. Shopkeepers depend on delivery of raw materials and manufactured parts from other shops, who got theirs from others, who got theirs . . . ultimately from the first shop, yesterday. Each action provokes the next, regularly following after a few hours' delay and triggering the next act in the normal sequence of civic and industrial life. Might we find that this sequence of time-delayed cues, leading in a circle out of long and well-established habit, still does so even in the absence of strict forcing by sunrise and sunset and mechanical clocks? In an analogous way, might the biochemical economy of each living cell have also incidentally internalized the habits forced on them a trillion times already without remission?

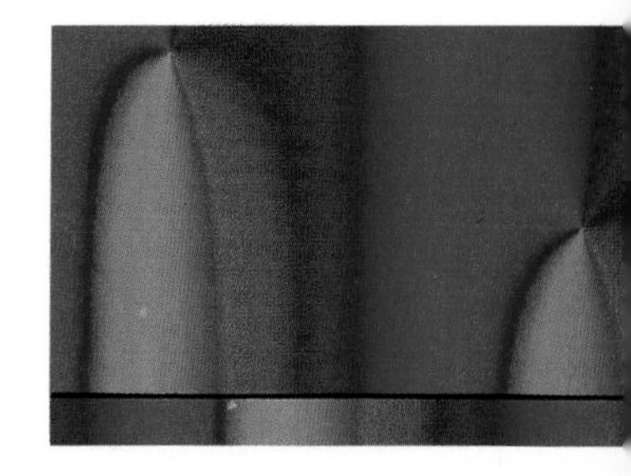

Third Dream: Inconstancy Is Inevitable

This vision starts not with an engineering goal, nor with a forced cycle, but with steady operation of cell chemistry, supposing that it is stable in a constant environment. But there is a difficulty with steady operation. The more complex the reactions, and the more regulatory coupling there is between reactions, the less is the chance that equilibrium will be stable. In other words, very few mutants of complex regulated pathways are capable of monotonously steady operation. Departure from monotonous steadiness need not lead to regular oscillations, but fluctuating mutants must be expected in overwhelming abundance over mutants whose equilibrium is well regulated.[4]

From this perspective, one might anticipate a diversity of rhythmic "mechanisms" as great as the diversity of schemes for feedback regulation—even within a single cell. Those that happen to conform well to regularities of the environment, such as the day/night cycle, might be exploited for some selective advantage. There may have been many independent evolutionary starts on this kind of internalization of diurnal rhythmicity, and no one had to win out. Moreover, with several independently competent systems oscillating within each cell, the clock is fail-safe, so each part is free to mutate without imperiling the function collectively subserved. Thus, no aspect of mechanism need be conserved during evolution. If circadian clocks arose in this way, they may have far less in common from the viewpoint of the biochemist than they do from the viewpoint of the naturalist or the theorist of dynamical systems.

Inconstancy, if bounded, may even be desirable in itself, especially in a constant environment. Cells commonly accommodate conflicting biochemical processes by compartmentation in space, but another perfectly respectable option is to compartmentalize in time, as we all do in our private lives. Why try to do everything at once? Sitting down and standing up at the same time accomplishes neither. Steady pressure on a nail accomplishes less than a rainfall of sharp hammer blows. Cycles of freeze and thaw can work wonders with a roadbed or, over a longer time, with a mountain. In a sailboat you can advance directly upwind by tacking back and forth, an impossibility for a boat with constant rudder. A steady background level of sexual tension doesn't motivate reproduction like the familiar buildup to a sudden release. It is better to sleep and wake at separate times than to groggily intermingle the two.

So we invent routines. Routines often lead back to the starting situation, so that we can do it again. Doing it again, we have set up a rhythm, for example, the rhythmic alternations of breathing, or those of movement in walking, running, swimming, or flying, or the cycles of rotating or reciprocating machinery. Realizing that vast improvements of efficiency can be derived from electrical reciprocation, Tesla and Steinmetz persuaded the nation to adopt an alternating current standard, replacing the original DC power network. In chemical factories, reaction vessels are filled and emptied, heated and cooled to efficiently convert raw materials into useful products. This might be expected, in general; freedom to depart from constancy opens up opportunities to find a better way of doing things. In complicated reactions, excursions to extreme conditions may "allow every dog his day"; a cycle allows the temporal segregation of incompatible reactions. Simple organisms, even single cells, being little chemical factories, may also benefit from this device.

In short, there may be reasons why, even if organisms didn't live in a fluctuating external environment, they could benefit by a "do-it-yourself" approach to rhythmicity. Computer simulation of randomly constructed control systems even suggests that intense selection would be needed to *avoid* spontaneous fluctuation. None of this specifically implies regular fluctuations, nor indeed is regularity needed for reaping many of the benefits of variability. But having started clock evolution (in our imaginations) by simply not selecting against the inevitable instabilities of monotony, we may now turn to the rhythmic environment for some refining selective pressures. Fluctuations that do happen to occur regularly, and at frequencies that closely match dominant regularities of the natural world, lend themselves to special uses.

We might expect, for example, the evolution of a capacity to sense the timing of environmental rhythms, and get in step with them. If the desert is hot in the afternoon, well, just set your internal clock so that your active, awake time is in the cool of the evening. If you need light for energy, arrange that your preparedness to absorb light reaches its daily maximum just before sunrise. If predators are on the prowl when you can't see at night, well, then arrange to be sleepily curled up in your burrow during those dangerous hours. In such arrangements, we find the most conspicuous benefits of circadian rhythmicity.

Must One of Them Be Right?

These three "dreams" are not mutually exclusive. All can be fleshed out in hypothetical details compatible with the facts of cued resetting. Rather than suggesting specific mechanisms, they suggest three distinct principles that might have contributed to the origin of circadian clocks, each dominating under different conditions. Did clocks of different sorts evolve under different conditions? We don't yet know, and it is hard to guess how we might find out; patterns of timing leave very few fossils.

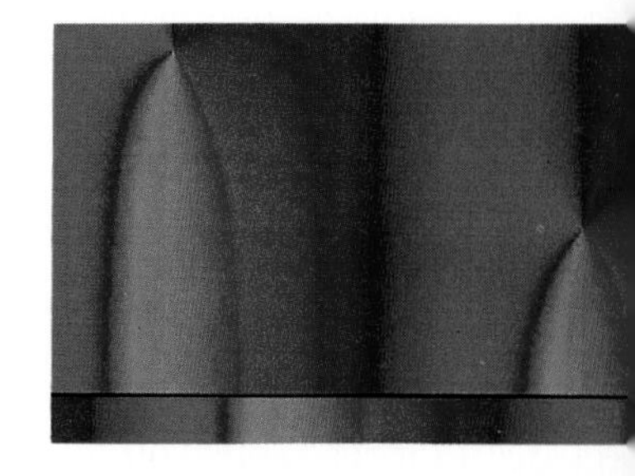

The timing of chemical reactions is periodically organized in this thin layer of liquid. Waves radiate from a pair of mirror-image phase singularities.

Chapter Nine

The Shape of the Future

Physics is mathematical not because we know so much about the world, but because we know so little: it is only its mathematical *properties that we have discovered.*

BERTRAND RUSSELL, *AN OUTLINE OF PHILOSOPHY*

Mathematically flavored principles have a way of reappearing in unexpected contexts. The topological principles of phase resetting in circadian clocks do indeed reappear, even in contexts that could be considerably more important than the original one.[1] An example of obvious practical importance was touched on in Chapter 5. The pacemaker node of the heart is a biophysical oscillator, subject to neural resetting stimuli. It turns out to reset in odd and even styles, connected together in a screw-shaped resetting surface, with a phase singularity. Another resettable oscillator of interest to neurobiologists governs breathing and perhaps the mysteriously abrupt cessation of breathing called sudden infant death.

Respiration

Is breathing regulated by a biological clock? It is possible to consider the question abstractly, paying no attention to what little is known of neural mechanisms in the brain stem. Can the normal rhythm of respiration be phase reset by neural impulses that alter inspiration and expiration? Are the phase resetting curves odd type for small stimuli and even type for larger stimuli? If so, is there a phase singularity? David Paydarfar and Frederic Eldridge asked these questions in recent experiments at the University of North Carolina Medical School.[2] They used adult cats, not infant humans, but the motivation is obvious: in sudden infant death, breathing stops for no apparent reason; could this have anything to do with phase singularities?

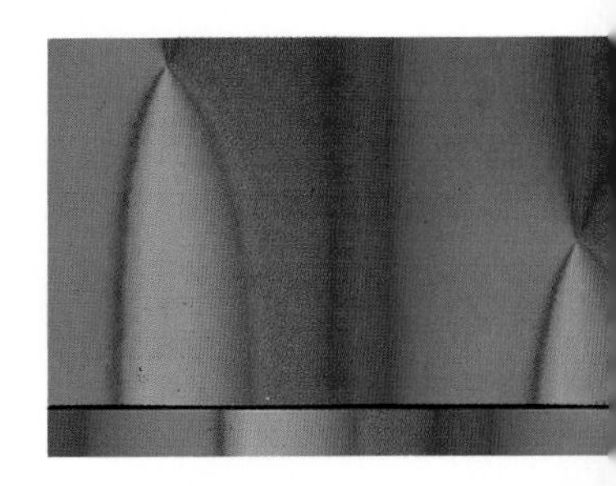

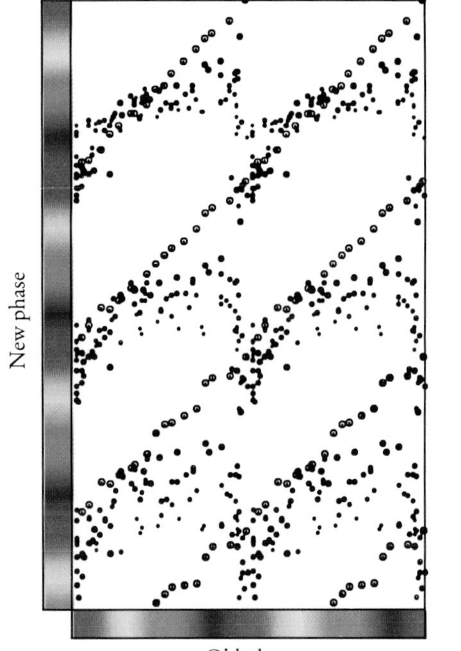

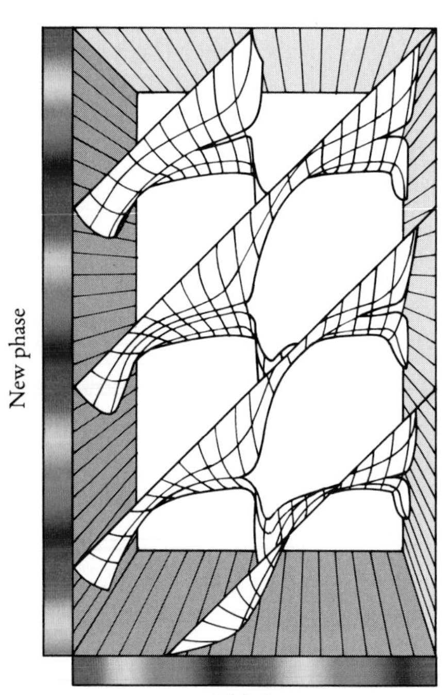

Above: Time crystal of the cat's respiratory oscillation, reset by neural impulses of varying duration. Each dot represents the beginning of an inspiration. Stimulus duration is 0 in the foreground (largest dots) to 2 seconds in the background (smallest dots). The standard cycle of breathing, and thus one cycle of phase, is about 5 seconds. The data are doubleplotted horizontally. *Below*: A contour map of new phase for the time crystal.

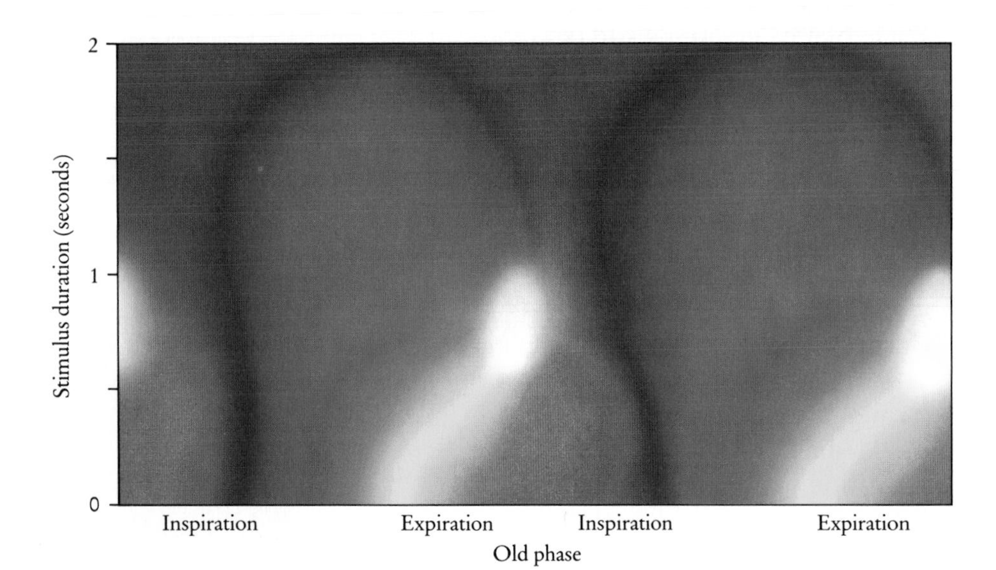

A definitive answer to this ultimate question is not yet available, but the experiments of Paydarfar and colleagues answer a clear yes to the preliminary questions. They stimulated the superior laryngeal nerve, a normal route for inhibitory impulses that shorten inspiration or prolong expiration. Varying the time of onset and the stimulus duration, they monitored the timing of activity in the phrenic nerve between the brainstem and the diaphragm. Reproducible phase resetting was obtained, in the odd pattern for stimuli less than half a second in duration, and in the even pattern for stimuli of a second or more. Plotted three-dimensionally, their data reveal a standard time crystal: a screw surface bounded by a helical border in each unit cell, all skewered on a singular axis. The corresponding color-coded contour map shows a color wheel surrounding a gray region of uncertainty. This singularity consists of the responses to stimuli of intermediate strength (about ¾ second) applied just before inspiration was expected. Such stimuli proved not to terminate breathing in the mature cat: the oscillator may have been forced to near equilibrium by near-singular stimuli, but it promptly resumed cycling at an unpredictable phase. Similar results were also obtained by using a facilitatory stimulus to the midbrain. If this means that equilibrium is very unstable in the mature, breathing animal, then the unborn animal's immature respiratory oscillator might be described as having a stable equilibrium that becomes unstable at about the time when breathing becomes spontaneous. Perhaps in some infants it is not yet very unstable. Then the rare accident of a singular stimulus might result in dangerously prolonged vacillation before the rhythm redevelops. Only a repeat of the same experiments with newborn animals will reveal whether it is sometimes too prolonged.

Nerve Cells Have Neighbors

Neurobiological oscillators such as the respiratory center or the heart's pacemaker consist of a great many distinct cells, each of which may be an oscillator in itself, yet we are accustomed to think of the entire mechanism as though it were a single neural pacemaker. Is mutual synchrony in fact flawless, even during the response to a stimulus? If not, as seems most likely, then the system's collective behavior may differ qualitatively from the behavior of a single individual, as we saw in Chapter 6. Fifteen years ago, a limit-cycle oscillation (and so all the implicit resetting behavior) was shown to result from mutual coupling of myriad model cells, each of which caricatures a neural pacemaker in such a crude way that its resetting curves are discontinuous and it has no equi-

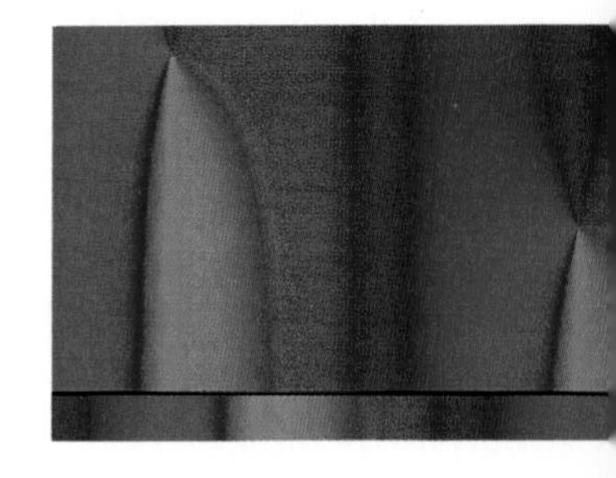

librium state.[3] A numerical model of the circadian pacemaker of the vertebrate brain has been built of such cellular units, none of which exhibit even resetting or a singularity; yet the aggregate of simulated cells has even resetting, a screw-shaped resetting surface, and a phase singularity.[4] It is clearly necessary to understand oscillator behavior in its natural context, in relation to the activities of neighboring cells.

This geometrical context has been uniformly absent from discussion of phase resetting in circadian clocks, perhaps because no one has yet monitored the spatial pattern of timing after a resetting stimulus in any circadian system. In contrast, spatial arrangements of timing are already the central focus of study in several other experimental systems of interest to physiologists. The pinwheel experiment provides a way of anticipating how resetting stimuli may affect such spatial patterns.

Rotating Waves

By systematically laying out the components of a time-crystal measurement in the format of a pinwheel experiment, we saw that phase resetting can be expected to alter the arrangement of timing in space (see page 96). In particular, a medium composed of clocks individually capable of even resetting may be provoked into rotation. This can be seen by flipping the pages of this book while watching the cycling of the color-coded phase at each point in the pinwheel experiment illustrated in the lower right corner. If clocks in their vulnerable phase fall within the scope of a stimulus whose strength straddles the critical level, then a phase singularity is implanted in the medium. Clocks nearby are so reset that a circular gradient of timing appears around the singularity. If neighboring points in this array of clocks are not connected, then a rotating wave persists until another stimulus intervenes or until all local clocks drift hopelessly out of synchrony with their neighbors. On the other hand, if neighboring cells are appropriately coupled, then local synchrony can be maintained and the vortex pattern can persist.

In computer simulations of idealized media, this is a common result, and the period of rotation shortens to nearly the minimum interval required for local recovery of excitability after each pulse. What about real media? Among circadian systems, a natural candidate for such investigations is a suspension of weakly interacting *Gonyaulax* cells.

Imagine an experiment using the pinwheel protocol on some dark beach deep inside a marine cave. The clock in *Gonyaulax* is paralyzed by the continuous glare of our Coleman lantern while we paint cells uniformly over a rectangular area of sand. Then they are released to begin cycling, column by column as a shadow slowly traverses the rectangle from east to west. After 23 hours, the eclipse is total and we have a gradient of phase from west (just started, at the shadow's edge) to east (started 23 hours ago). A flashlight beam briefly swept along the north side now exposes northern clocks at all old phases, depending on their east-west location. Clocks more to the south are less exposed. Thus, in one shot, all combinations of old phase and stimulus size appear in orderly arrangement on the rectangle. The results will be observed as a locus of synchronous glow (an isochron), moving like a wave from cell to cell across the rectangle.

Soon after the flash, a spot of intense blue bioluminescence appears somewhere in the middle of the rectangle, lingering for several hours. As it later fades and vanishes, a wave of softer glow appears, pirouetting once every 23 hours around that little patch of perpetual darkness. Here we find the singularity, where all time zones converge. The pinwheel isochron wave pivots about these cells as it advances from time zone to time zone.

The pinwheel experiment as actually performed in separate vials of *Gonyaulax* leaves each population of cells free of its neighbors, so the singular population is free to return to the usual cycle. That is not quite so in this idealized version free of vial walls. In this form, the pinwheel experiment suggests a way to achieve, contain, and study the otherwise volatile singular state in a tissue of connected clocks. The new ingredient is *connectedness.* If the adjacent "vials" of a pinwheel experiment are contiguous and interacting to maintain mutual synchrony as far as possible, then vials in the middle of a circle of phase (including the singular vial) have no choice but to remain intermediate between all phases. The singularity cannot get out as long as the surrounding cells remain on the usual cycle at definable phase. The singularity is entrapped for leisurely laboratory observation. Because *Gonyaulax* cells in the singular state show no strong tendency to return to normal oscillation anyway, this entrapment does not require particularly strong coupling between cells. *Gonyaulax* would thus offer a favorable opportunity to study gradients of circadian timing, and particularly to contrast the biochemistry of cells in this zero-amplitude state with that of normally cycling neighbors.

Circumstances for coupling would be more favorable if not even membranes separated cells. For example, the breadmold *Neurospora crassa* grows as a thin film of interconnected filaments. There are no cells in the usual sense; each filament has partitions along its length, but the partitions have holes almost large enough to pass organelles. In the

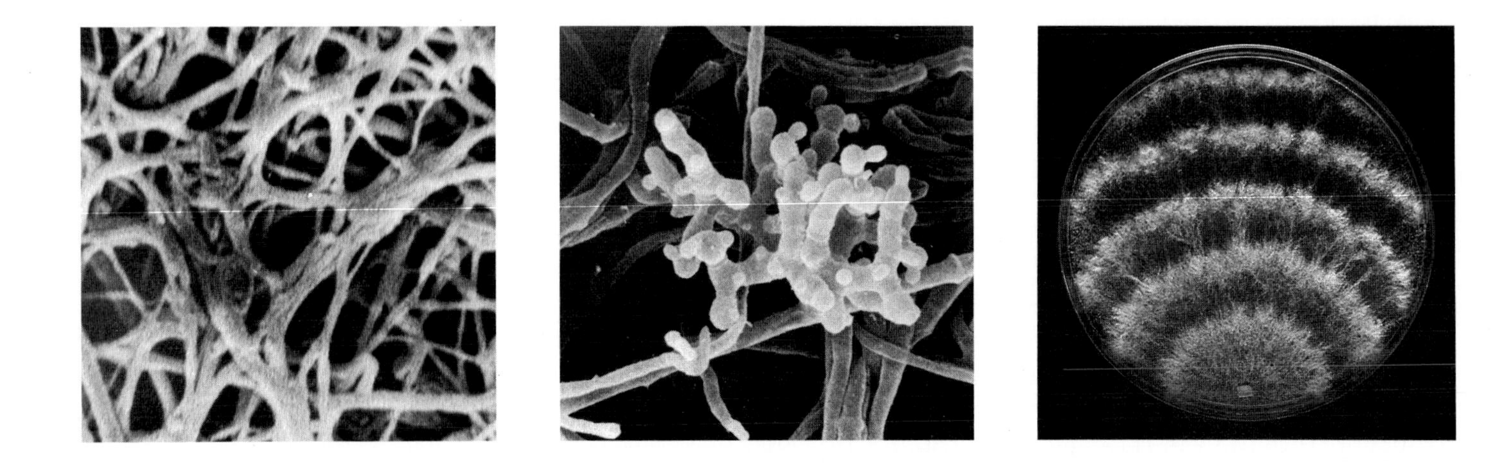

Above left: The breadmold *Neurospora* grows as a mat of interconnecting filaments, each only a few thousandths of a millimeter across. *Center*: From time to time, at circadian intervals, newly formed filaments on the growth frontier become predestined to develop spores on aerial threads. *Above right*: On a larger scale, a fuzz of such threads marks past positions of the growth frontier at intervals of about 2 centimeters. Each interval corresponds to 22 hours' expansion of the colony from the inoculum shown at the bottom.

dark, every part of this web of cytoplasmic filaments keeps time with a 22-hour period: if it is allowed to proliferate, at the right time in each cycle the new filaments will include spore-bearing *conidia*. Whatever this biochemical clock may consist of, it continues its circadian oscillation for many days at least and responds with even resetting to a light pulse. Steady blue light arrests the clock's progress. Normally the whole mat is synchronous. But subjected to a moving shadow that cuts off exposure to dim blue light, the clocks in a mat can be set in time zones along a gradient that spans whole cycles of circadian phase. Subjected to a transversely graded pulse of light, all these clocks are simultaneously reset as in any other pinwheel experiment, presumably entrapping a phase singularity somewhere in the interior. What happens next is not yet known; in all trials up to the present, the mat stopped growing at about this time.[5]

In the related fungus *Nectria cinnabarina,* circular gradients of phase form spontaneously at the outset of growth. As in *Neurospora,* phase here is indicated by the timing of spore deposition (shiny yellow ridges) along the frontier of growth, but in 16-hour cycles. Circular gradients of timing are revealed by the spiral patterns of deposited ridges; the singularity remains entrapped at the center for at least a week before escaping, as revealed by a change of pattern. No one yet knows how to create such a situation deliberately by controlled phase resetting in *Nectria,* or how to assay the phase anyplace but on the moving fringe of the growing mat. A more congenial experimental subject—perhaps a tissue of coupled circadian clocks governing the daily rhythm of stomatal opening in a plant leaf—will someday reveal the unstable singularity lashing about within the cage formed by a circular phase gradient.

A suspension of yeast cells would seem a natural candidate for a short-period oscillator, since it demonstrably responds to diverse stimuli in the even style, exhibits phase singularities, and can be arranged as a two-dimensional film of chemically coupled cells. Stimulation graded across a gradient of phase as in the pinwheel experiment might thus expose the violently unstable steady state of oscillating sugar metabolism to biochemical study in experiments as yet untried. It has been tried in a nonliving chemical caricature of biological clocks called the Belousov-Zhabotinsky reaction.

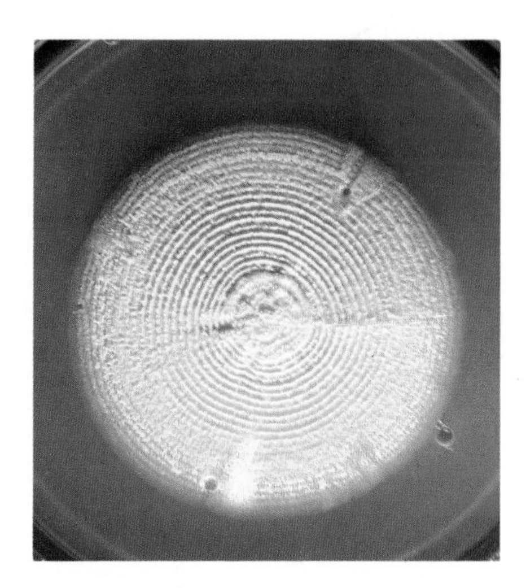

A disk of the fungus *Nectria cinnabarina* has grown from spores deposited in the center of a petri dish. Every sixteenth hour the disk's growing edge initiated development of spores: the yellow ridges seen at intervals of about 1 millimeter.

A Nonliving Chemical Oscillator

In Moscow about 1950, chemist Boris Belousov sought to create a simplified caricature of the Krebs cycle (citric acid cycle), a sequence of reactions central to the metabolism of all oxygen-using cells. He failed, in that the only compounds his final recipe had in common with its biochemical prototype were water and citric acid. Nonetheless he found something so unexpected that his critics branded it incredible and refused him publication in the chemistry journals. Only ten years after his death was Belousov honored with the 1980 Lenin Prize, along with Anatoly Zhabotinsky, Albert Zaikin, Valentin Krinsky, and Genrik Ivanitsky, all for work involving the Belousov reaction. What was so surprising that it had to wait thirty years for recognition?

The surprise was that Belousov's simple solution of five commonplace chemicals in water at room temperature did not react its way directly to a new equilibrium, but rather alternated between one composition and another with almost perfect regularity for up to a hundred cycles before finally exhausting one of the reactants. Such behavior was of course known in living systems but was not expected in this seemingly simple situation. This chemical oscillator responds to a pulse of ultraviolet light or a chemical stimulus by phase resetting in odd or even style, depending on stimulus strength.

In subsequent modifications of Belousov's recipe, the same reaction was found to respond to excitation much as nerve membrane does, and to propagate colorful waves of chemical activity that are in many ways analogous to the nerve impulse. In two dimensions, the solution promptly becomes organized in rhythmic wave patterns, every point alternating red/blue/red at a common period. Waves are gradients of phase. In some places, they circulate about a pivot—a phase singularity.

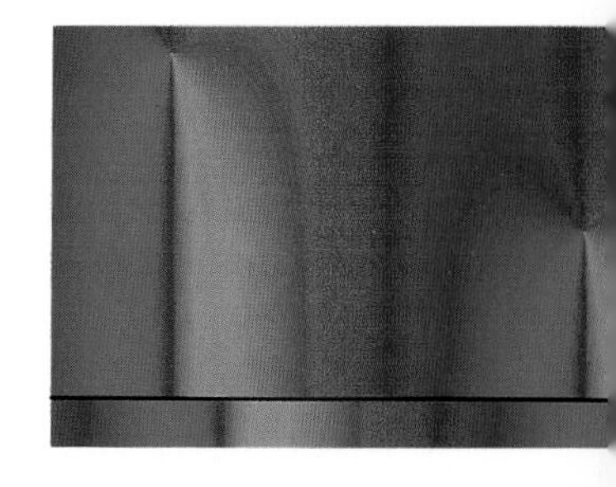

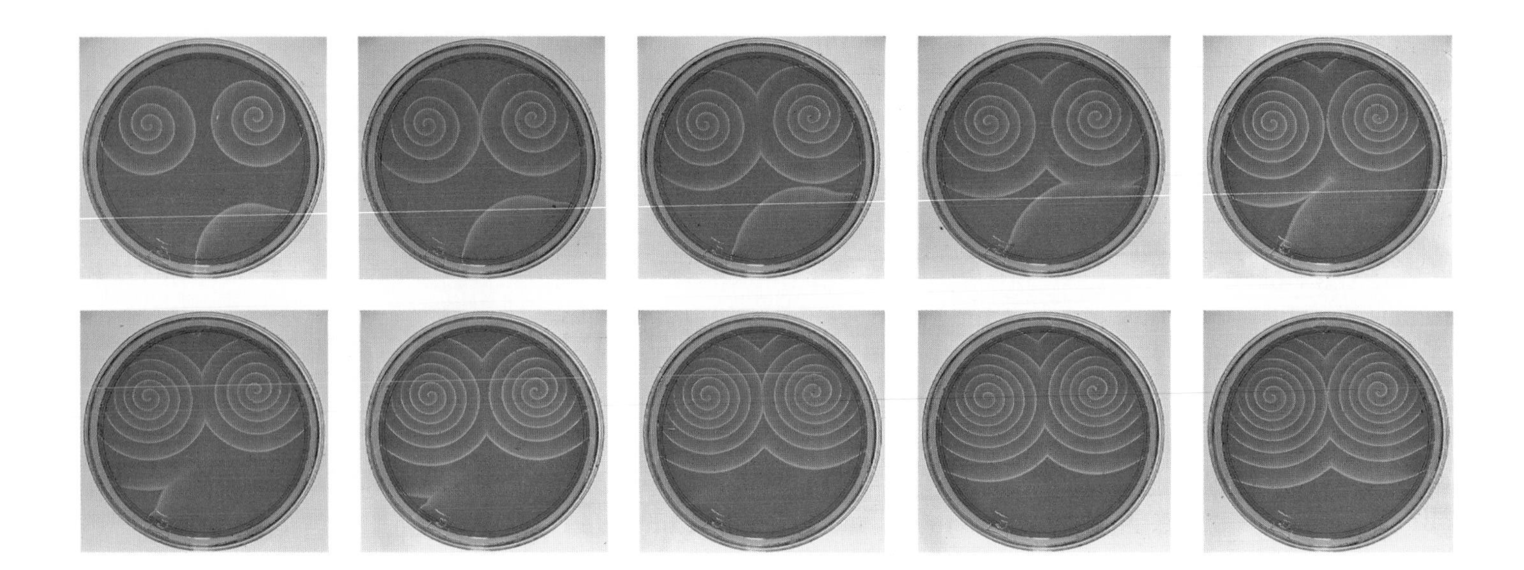

A sequence of snapshots ½ minute apart above a 1-mm layer of Belousov-Zhabotinsky reagent in a petri dish. The liquid is motionless, but a blue pulse of chemical activity is propagating almost radially from each of two centers. The wavefront is not a set of concentric rings, but a pair of mirror-symmetric spirals. Timing of the red/blue/red alternation throughout the dish is organized in circular phase gradients around each source. The dish is 47 millimeters across. The spiral rotates once per minute. The tiny circles are bubbles of carbon dioxide released as the organic substrate oxidizes.

To find out how you can perform this experiment yourself, see footnote 6.

Such a point can be created in the way that has become familiar: a gradient of stimulation (ultraviolet light) falling across a phase gradient (a wave) resets timing everywhere. If stimulus magnitudes are appropriate, a segment of wavefront is erased or created. Each endpoint is a phase singularity about which the wave then circulates, winding into a spiral.[6] This phenomenon is not just a quirk of exotic chemistry, but a principle of organization in rhythmic media.

A Phase Singularity in Cell Aggregation

A kind of cell called a social amoeba begins life as a solitary wanderer, then gathers with others to constitute a multicellular organism for purposes of reproduction. The gathering is organized by chemical waves propagating from cell to cell whenever enough amoebae have gathered into an area and run out of bacteria to eat. These waves typically have a period of about five minutes and emanate from a pacemaker cell whose spontaneous period is somewhat shorter than that of its neighbors (if they have matured to spontaneous oscillation at all). Such cells suspended in a water slurry still pulse periodically, releasing the pheromone cyclic AMP at regular intervals, or a little off set if stimulated by a pulse of cyclic AMP added to the suspension. Such a stimulus resets the phase of the spontaneous oscillation; a big enough stimulus resets in the

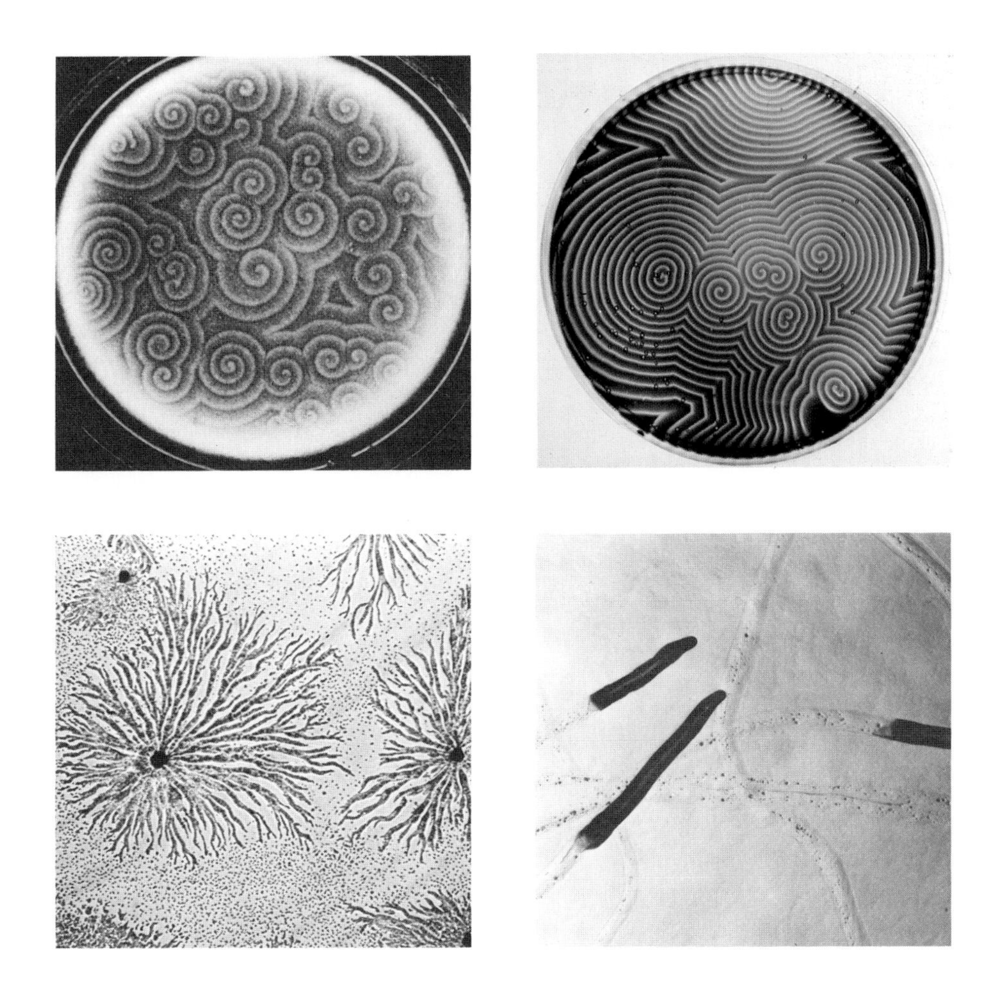

Above left: A monolayer of the slime mold *Dictyostelium discoideum* in a petri dish shows wave fronts of newly released cyclic AMP. *Above right*: The biochemical kinetics involved resembles that of the Belousov-Zhabotinsky reagent (shown here at the same magnification), as does the resulting organization of reactions in time and space. *Below left*: At a stage later than shown in the preceding figure, individual social amoebae of the slime mold *Dictyostelium discoideum* have linked into traffic streams converging toward several wave sources a few millimeters apart. *Below right*: Several hours later they have formed a simple organism 1 to 2 millimeters long, which crawls away to reproduce, still led by the pulsing wave source at its tip.

even pattern. If cells grow excitable but, for some reason, there are no pacemakers, propagating wavefronts of cyclic AMP arise anyway. They emanate from phase singularities in the spatial arrangement of timing, each being the center of a full-cycle gradient of phase. When a cell is hit by a wave of cyclic AMP, not only does it release its own cyclic AMP, but it also moves in the direction of the source. In this way cells gather and mutually adhere to form a snail-like *grex*. The finished grex crawls away and turns into dusty spores, restarting this ancient life cycle.

Cyclic waves in reacting media are of course not strictly two-dimensional: the liquid film has some depth, about 1 millimeter in the strip of

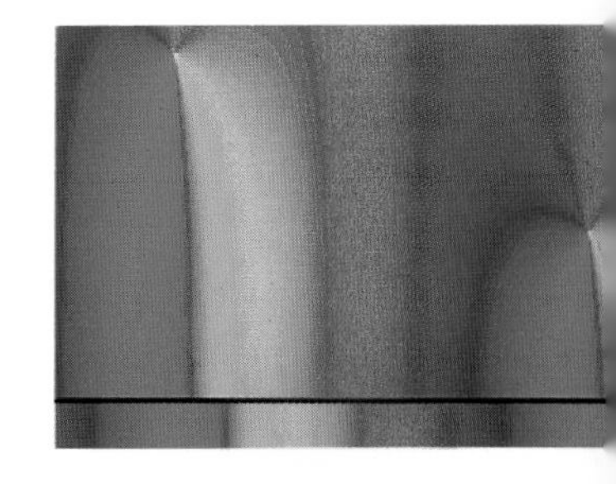

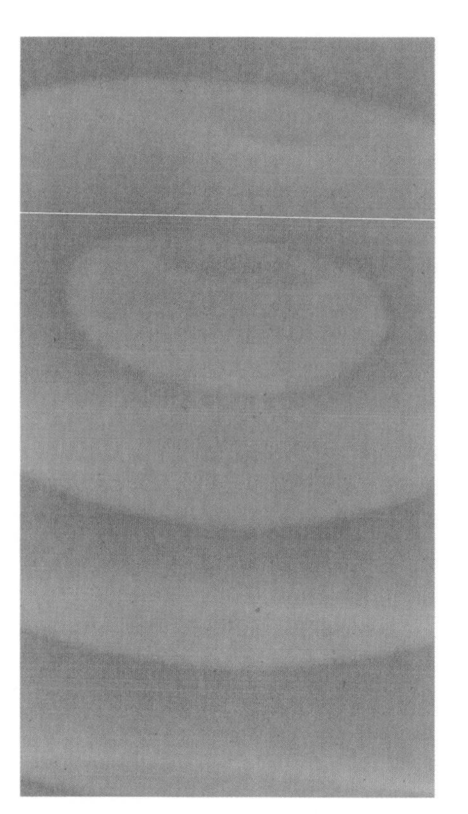

Left: A scroll ring photographed in a test tube, 10 mm wide, of Belousov-Zhabotinsky reagent. *Right*: A scroll ring by Cray computer graphics, with wave fronts cut away to expose interior anatomy; the singular ring, seen edge on, is red.

photographs on page 174. The phase singularity is thus not a pivot *point* but a 1-millimeter-long vertical axle around which the timing of reactions is sequenced through one full cycle. In thicker films, the singular axle is a filament threading its way from glass through liquid to air or back to glass, or else closing in a perfect ring. Around that filament revolves the three-dimensional analog of a spiral: a scroll-shaped wave. Photographed in frozen projection through a spacious layer of liquid, singular filaments seem more an inference than a direct observation. But watched through a low-power microscope, scroll rings clearly rotate in a fully three-dimensional volume of motionless liquid, as seen in the photo above. The crisper idealization beside it was photographed inside the memory of a supercomputer. It reveals the anatomy of scroll rings more clearly than any single photograph of the chemical reality.

The chemistry of the Belousov-Zhabotinsky reaction is now known in fairly complete detail. The physics of coupling between adjacent microscopic volumes (by molecular diffusion) is exactly understood. The mathematics of its organization in space and time reveals a variety of topologically distinct arrangements of ring-shaped singular filaments, like smoke rings or ring-shaped tornadoes, except that there is no physical movement. Each such organizing center sequences the reactions throughout a volume of liquid by radiating waves of chemical change at a characteristic short period. A volume of medium perfused with singularities trapped inside full cycles of chemical change is strictly unable to oscillate as a unit, but exhibits complicated high-frequency activity.

Spatial Disorganization of the Heartbeat

The equations of the Belousov-Zhabotinsky reaction strikingly resemble those used by electrophysiologists to quantify membrane excitability; and the spatial coupling terms in both sets of equations have the same mathematical form. Both media support oscillations and pulselike waves with the same peculiar characteristics.

The pacemaker node of the heart exhibits the electrical rhythmicity implicit in those equations. The timing of neural and cardiac pacemak-

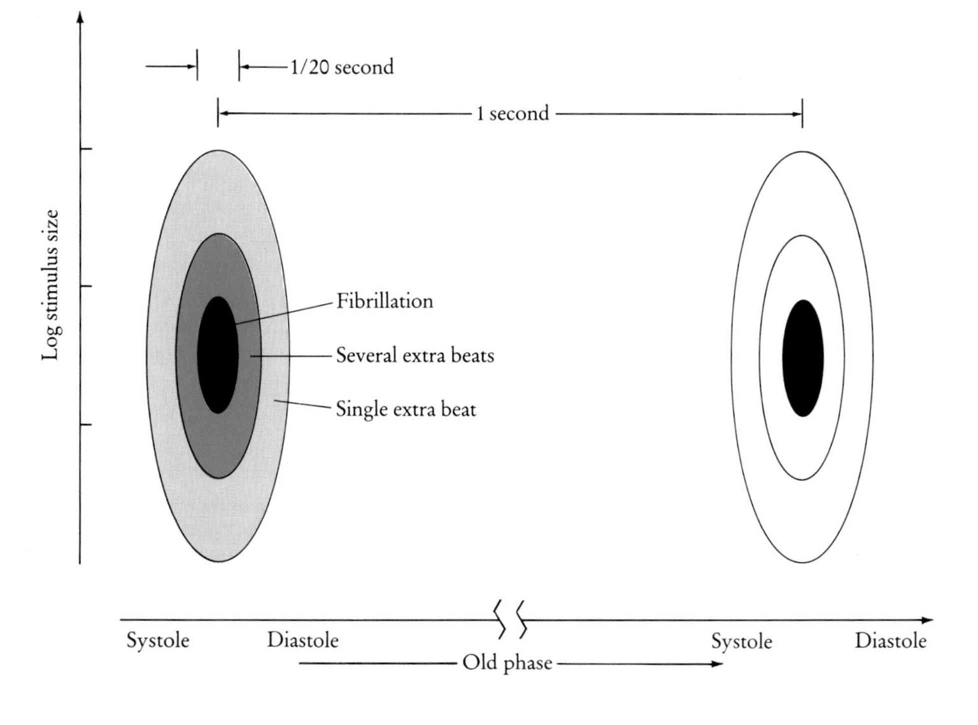

On the same coordinates as former colored phase-resetting diagrams, this sketch neglects phase to show zones (shaded) in which the stimulus evokes one or several extra beats, or (black bull's-eye) continuing high-frequency activity (fibrillation). The window of vulnerability is less than 1/20 cycle wide and spans a roughly tenfold range of stimulus size.

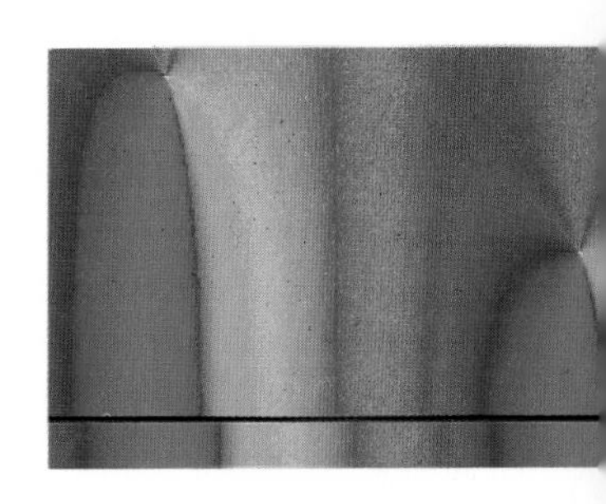

ers is typically adjustable by stimuli from nerve fibers. The reset inflicted by a single impulse depends on its size and on its timing. A decade ago, a search was undertaken for even resetting in such media. Many examples were uncovered; not much later, the predicted phase singularities were also sought and found. The principles pioneered by biologists studying circadian rhythms have been borrowed by cardiac physiologists to develop new interpretations of lethal arrhythmia in the heart muscle, and even have yielded new clinical procedures to terminate unwanted rhythms.

One class of arrhythmias induced by stimulating the heart in its vulnerable phase may consist of three-dimensional organizing centers in the thick left ventricular muscle. These would be visible on the surface or in thinner layers as mirror-image vortices of electrical activity rotating at nearly the minimum interval required for local recovery of excitability after each pulse. Such stable counter-rotating arrhythmias should be inducible at certain places in the heart muscle by stimuli such as electrical shocks of various intensities applied at various moments during the vulnerable interval. Plotted on a diagram of stimulus size versus phase of heartbeat such stimuli should occupy a delimited region surrounded on all sides by a halo of stimuli that only elicit a few quick extra beats; stimuli outside that bull's-eye should be safe.

These conjectures can be tested directly in the laboratory. At latest report,[7] they seem to realistically describe the onset of ventricular fibrillation, the lethal arrhythmia implicated in most cases of sudden cardiac death.

The study of cued resetting has led naturally from jet lag and phase singularities of the global time zones, to phase singularities in circadian rhythms that are not spatially organized, back to phase singularities embedded in geography, but now in three-dimensional media such as rhythmically active heart muscle. The next several years will almost certainly bring an answer to the question whether lethal activity in the heart muscle initially consists of phase singularities and waves rotating around them in recognizable simplicity. Because of the longer time scale of circadian experiments, decades may pass before the pertinence of these principles to human circadian physiology is clarified. Is it possible that the principles first deciphered by experiments on circadian clocks will in the end prove relevant to everything except human circadian rhythms? It seems unlikely. Discounting a half century of uncontrolled experiments in jet lag on commercial airliners, our body clocks have been accessible to observation for only about twenty years, and only in the past five years have temporal-isolation facilities been constructed in many research centers. Though no one has yet measured a phase-resetting curve on the human circadian clock, it seems clear that we must have one, and we would be unusual if even resetting is not in our repertoire.[8] No one has yet interpreted the uncatalogued arrhythmias of the human clock in the fashion made respectable by cardiologists, but it seems clear that there are disorders of phasing, arrhythmias, and other aberrations of circadian timing, all too evident in human insomnia, in social habits, and in both physiological and psychiatric illnesses. Shift work, jet travel, and space flight make them even more conspicuous. Progress will be slow; as Hermann Hesse says in *The Steppenwolf Treatise,* "Things are not so simple in life as in our thoughts, nor so rough and ready as in our poor idiotic language." But some new points of view have now been pioneered; there is too much precedent in electrophysiology, in cardiology, in physical chemistry, and in applied mathematics for aberrations of circadian temporal organization to forever resist analytical description, and interpretation, and remedy.

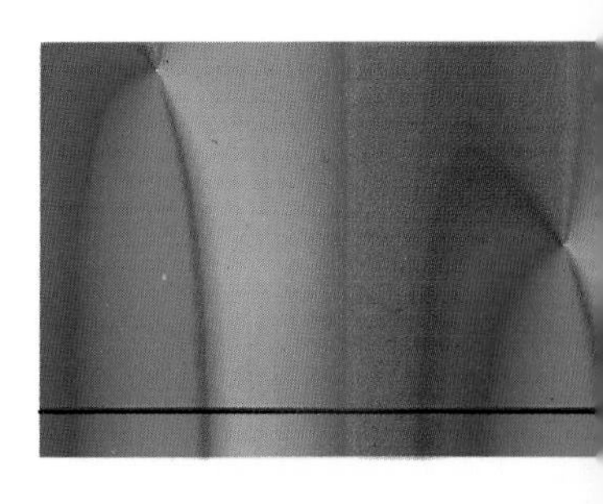

For Further Reading

In this book I have tried hard not to embroil the reader in more scholarly debate than is necessary, while still presenting a valid low-resolution picture of the theory and the tantalizing questions that await answers. Albert Einstein once remarked, "Everything should be made as simple as possible—but not simpler." In doing so, one runs the risk that proves fatal to almost all pornographic movies: by simplifying too much, they lose the essence. I hope my judgment does not offend. The following sources, in any case, provide lucid amplification and the context surrounding the few principles extracted for this book.

Bunning, E. (1973) *The Physiological Clock.* New York: Springer-Verlag. This thin paperback reprints and updates Bunning's classic gathering of what was known about circadian rhythms in the early 1960s. It remains a useful source.

R. A. Wever (1979) *The Circadian System of Man.* Heidelberg: Springer-Verlag. In a landmark monograph, Wever summarizes explorations of normal human sleep timing, core temperature rhythm, and so on in hundreds of young adult subjects of both sexes under conditions of temporal isolation and under diurnal forcing at various periods.

A. T. Winfree (1980) *The Geometry of Biological Time.* New York: Springer-Verlag. The first half presents the qualitative theory of phase control in simple clocks, attracting-cycle oscillators, and populations of such units. The second half presents a dozen experimental systems that illustrate (among other things) the phenomena anticipated or analyzed in the first half. A bibliography of 1400 research papers is indexed to the 458–page text.

M. C. Moore-Ede, F. M. Sulzman, and C. A. Fuller (1982) *The Clocks That Time Us.* Cambridge, Mass.: Harvard University Press. Lively and lucid writing distinguish this summary of circadian experiments (mostly on primates and other mammals) up to about 1981.

M. C. Moore-Ede, C. A. Czeisler, and G. S. Richardson (1983) "Circadian Timekeeping in Health and Disease. Part I: Basic Properties of Circadian Pacemakers." *New England Journal of Medicine,* 309:469–476. "Part II: Clinical Implications of Circadian Rhythmicity." *New England Journal of Medicine,* 309:530–536. A provocative plea to the medical profession to make more use of new understanding of circadian temporal organization.

A. Reinberg and M. Smolensky (1983) *Biological Rhythms and Medicine.* New York: Springer-Verlag. The emphasis here is on tissue and cell rhythms in humans, other primates, and laboratory rodents. Topics include circadian timing of allergic responses, and of cell division in connection with cancer chemotherapy.

M. C. Moore-Ede and C. A. Czeisler, eds. (1984) *Mathematical Modeling of Circadian Systems.* New York: Raven Press. This is a 1981 symposium proceedings, largely updating the themes of Wever's monograph, two years later.

A. T. Winfree (1987) *When Time Breaks Down: The Three-dimensional Dynamics of Electrochemical Waves and Cardiac Arrhythmias.* Princeton, N.J.: Princeton University Press. This somewhat more technical presentation continues where *The Timing of Biological Clocks* leaves off. After an introductory chapter, *When Time Breaks Down* dwells mostly on spatial patterns surrounding phase singularities in excitable media such as the Belousov-Zhabotinsky reagent and heart muscle.

If you like geometry, I recommend to your attention the details of the visual theory of dynamics sketched in Chapter 8. For inspirational reading start with FitzHugh (1960, 1961) on neural pacemakers and Abraham and Shaw (1982, 1984, 1985) on topological visions:

R. FitzHugh (1960) "Thresholds and Plateaus in the Hodgkin-Huxley Equations." *Journal of General Physiology,* 43:867–896.

——— (1961) "Impulses and Physiological States in Theoretical Models of Nerve Membrane." *Biophysical Journal,* 1:445–466. Using the graphical language of state spaces, trajectories, and limit cycles FitzHugh paints an image of neural dynamics that provides the reader with a deeper, intuitive understanding of the complex responses of pacemakers to stimuli.

R. H. Abraham and C. D. Shaw (1982) *Dynamics: The Geometry of Behavior. Part 1: Periodic Behavior* (1982). *Part 2: Chaotic Behavior*

(1984). *Part 3: Global Behavior* (1985). Santa Cruz, Calif.: Aerial Press. These three paperbacks are unique for their essentially pictorial presentation of subtle topology by four-color cartoons.

Someday when laboratory investigations of circadian physiology are a little more mature, we will know the chemical names of quantities that are only coordinate directions in present-day theory. Then the phenomenological approach of this book will seem very Spartan, like genetics before molecular biology. But timing and dynamics and behavior will never be adequately represented in words and chemical names. The story will again be told in the geometrical language of its origin: the language of attracting cycles in state spaces, but with specific biochemical and biophysical labels on coordinates that have remained only abstractions for over twenty years.

References

Chapter One

1. L. Carroll (1971) *The Rectory Umbrella and Mischmasch.* New York: Dover Publ., pp. 31–32.

2. D. K. Winstead, B. D. Schwartz, and W. E. Bertrand (1981) "Biorhythms: Fact or Superstition?" *American Journal of Psychiatry,* 138(9): 1188–1192.

   D. F. Kripke, H. Yelverton, and Zd. Kripke (1979) "Biorhythm is Bio-Nonsense." *American Biology Teacher,* 41: 108–109.

   J. W. Shaffer, C. W. Schmidt, Jr., H. I. Zlotowitz, et al. (1978) "Biorhythms and Highway Crashes: Are They Related?" *Archives of General Psychiatry,* 35: 41–46.

   J. H. Wolcott, R. R. McMeekin, R. E. Burgin, et al. (1977) "Correlation of Occurrence of Aircraft Accidents with Biorhythmic Criticality and Cycle Phase in U.S. Air Force, U.S. Army, and Civil Aviation Pilots." *Aviation, Space, and Environmental Medicine,* 48: 976–983.

   See extensive bibliographies in:

   K. E. Klein and H. M. Wegmann (1979) "Circadian Rhythms of Human Performance and Resistance: Operational Aspects." Nuilly-sur-Seine: NATO-AGARD publication AG-105.

   D. C. Holley, C. M. Winget, and C. M. DeRoshia (1981) "Effects of Circadian Rhythm Phase Alteration on Physiological and Psychological Variables: Implications to Pilot Performance." NASA Technical Memorandum 81277.

3. W. Whewell (1836) "Researches on the Tides." *Philosophical Transactions of the Royal Society of London,* 126: 289–307.

4. E. Schwiderski (1983) "Atlas of Ocean Tidal Charts and Maps. Part I: The Semidiurnal Principal Lunar Tide M_2." *Marine Geodesy,* 6(3–4): 219–265.

Chapter Two

1. J. Aschoff (1965) "Circadian Rhythms in Man." *Science,* 148: 1427–1432.

2. M. C. Moore-Ede, F. M. Sulzman, and C. A. Fuller (1982) *The Clocks That Time Us.* Cambridge, Mass.: Harvard University Press.

   M. C. Moore-Ede, C. A. Czeisler, and G. S. Richardson (1983) "Circadian Timekeeping in Health and Disease. Part I: Basic Properties of Circadian Pacemakers." *New England Journal of Medicine,* 309: 469–476. "Part II: Clinical Implications of Circadian Rhythmicity." *New England Journal of Medicine,* 309: 530–536.

3. E. Haus, F. Halberg, L. E. Scheving, J. E. Pauly, et al. (1972) "Increased Tolerance of Leukemic Mice to Arabinosyl Cytosine with Schedule Adjusted to Circadian System." *Science,* 177: 80–82.

   E. Haus, F. Halberg, J. F. Kuhl, and D. J. Lakatua (1974) "Chronopharmacology in Animals." *Chronobiologia,* 1: 122–156.

   A. Reinberg and M. Smolensky (1983) "Biological Rhythms and Medicine," New York: Springer-Verlag, Chapter 4. This text describes more recent experiments that substantially extend the results of the pioneering studies cited above.

4. D. Kripke (1984) "Critical Interval Hypothesis for Depression." *Chronobiology International,* 1(1): 73–81.

   ——— (1983) "Phase-Advance Theories for Affective Illnesses." In *Circadian Rhythms in Psychiatry,* edited by T. A. Wehr and F. K. Goodwin, pp. 41–69. Pacific Grove, Calif.: Boxwood Press.

   T. A. Wehr and F. K. Goodwin (1981) "Biological Rhythms and Psychiatry." In *American Handbook of Psychiatry,* Vol. 7, Chapter 3, pp. 46–74.

5. T. J. Savides, F. Messin, C. Senger, and D. F. Kripke (1985) "Natural Light Exposure of Young Adults." *Sleep Research,* 14: 310.

N. Okudaira, D. F. Kripke, and J. B. Webster (1983) "Naturalistic Studies of Human Light Exposure." *American Journal of Physiology,* 245: R613–R615.

6. C. P. Richter (1968) "Inherent 24-hour and Lunar Clocks of a Primate—The Squirrel Monkey." *Communications in Behavioral Biology,* 1: 305–332.

7. L. E. M. Miles, D. M. Raynal, and M. R. Wilson (1977) "Blind Man Living in Normal Society Has Circadian Rhythms of 24.9 Hours." *Science,* 198: 421–423.

8. C. P. Kokkoris, E. D. Weitzman, C. P. Pollak, A. J. Spielman, C. A. Czeisler, and H. Bradlow (1978) "Long-term Ambulatory Temperature Monitoring in a Subject with a Hypernychthermal Sleep-Wake Cycle Disturbance." *Sleep,* 1: 177–190.

9. A. L. Weber, M. S. Cary, N. Connor, and P. Keyes (1980) "Human Non-24-Hour Sleep-Wake Cycles in an Everyday Environment." *Sleep,* 2: 347–354.

10. M. Jouvet, J. Mouret, G. Chouvet, and M. Siffre (1974) "Toward a 48-hour Day: Experimental Bicircadian Rhythm in Man." In *The Neurosciences Third Study Program,* edited by F. Schmitt and F. Worden, pp. 491–497. Cambridge, Mass.: MIT Press.

11. A. T. Winfree (1982) "Circadian Timing of Sleep and Wakefulness in Men and Women." *American Journal of Physiology,* 243: R193–R204.

    ——— (1982) "Human Body Clocks and the Timing of Sleep." *Nature,* 297: 23–27.

    ——— (1983) "The Impact of a Circadian Clock on the Timing of Human Sleep." *American Journal of Physiology,* 245: R497–R504.

    ——— (1984) "Exploratory Data Analysis: Published Records of Uncued Human Sleep and Waking." In *Mathematical Modeling of Circadian Systems* (Hyannis 1981), edited by M. C. Moore-Ede and C. A. Czeisler, pp. 187–200. New York: Raven Press.

12. J. Foret and G. Lantin (1972) "The Sleep of Train Drivers: An Example of the Effects of Irregular Work Schedules on Sleep." In *Aspects of Human Efficiency,* edited by W. P. Colquhoun, pp. 273–282. London: English Universities Press.

13. C. A. Czeisler (1978) "Human Circadian Physiology: Internal Organization of Temperature, Sleep-Wake, and Neuroendocrine Rhythms." Ph.D. dissertation, Stanford University.

    C. A. Czeisler, E. D. Weitzman, M. C. Moore-Ede, J. C. Zimmerman, and R. G. Knauer (1980) "Human Sleep: Its Duration and Organization Depend on Its Circadian Phase." *Science,* 210: 1264–1267.

14. J. Zulley, R. Wever, and J. Aschoff (1981) "The Dependence of Onset and Duration of Sleep on the Circadian Rhythm of Rectal Temperature." *Pflugers Archiv,* 391: 314–318.

15. See note 11 above.

16. S. H. Strogatz, R. E. Kronauer, and C. A. Czeisler (1986) "Circadian Regulation Dominates Homeostatic Control of Sleep Length and Prior Wake Length in Man." *Sleep,* in press.

    S. H. Strogatz and R. E. Kronauer (1986) "The Rules of the Human Circadian Sleep-Wake Cycle." (in press)

Chapter Three

1. A. J. Lewy, T. A. Wehr, F. K. Goodwin, and D. A. Newsome (1980) "Light Suppresses Melatonin Secretion in Humans." *Science,* 210: 1267–1269.

   A. J. Lewy (1983) "Effects of Light on Human Melatonin Production and the Human Circadian System." *Progress in Neuro-psychopharmacology and Biological Psychiatry,* 7: 551–556.

   A. J. Lewy, R. A. Sack, and C. L. Singer (1984) "Assessment and Treatment of Chronobiologic Disorders Using Plasma Melatonin Levels and Bright Light Exposure: The Clock-Gate Model and the Phase Response Curve." *Psychopharmacology Bulletin,* 20(3): 561–565.

2. M. Menaker, J. E. Takahashi, and A. Eskin (1978) "The Physiology of Circadian Pacemakers." *Annual Reviews of Physiology,* 40: 501–526.

   N. H. Zimmerman and M. Menaker (1979) "The Pineal Gland: The Pacemaker Within the Circadian System of the House Sparrow." *Proceedings of the National Academy of Sciences USA,* 76: 999–1003.

   J. Takahashi and M. Zatz (1982) "Regulation of Circadian Rhythmicity." *Science,* 217: 1104–1110.

3. J. Redman, S. Armstrong, and K. T. Ng (1983) "Free-running Activity Rhythms in the Rat: Entrainment by Melatonin." *Science,* 219: 1089–1092.

   G. E. Pickard and F. W. Turek (1983) "The Suprachiasmatic Nuclei: Two Circadian Clocks?" *Brain Research,* 268: 201–210.

   F. C. Davis and R. A. Gorski (1984) "Unilateral Lesions of the Hamster Suprachiasmatic Nuclei:

Evidence for Redundant Control of Circadian Rhythms." *Journal of Comparative Physiology,* A154: 221–232.

4. P. E. Mullen, C. Linsell, R. E. Silman, R. Edward, et al. (1978) "The Human Pineal: New Approaches and Prospects." *Journal of Psychosomatic Research,* 22: 357–376.

   F. W. Turek and S. Losee-Olsen (1986) "A Benzodiazepine used in the Treatment of Insomnia Phase Shifts the Mammalian Circadian Clock." *Nature,* 321: 167–168.

   T. A. Wehr and F. K. Goodwin (1981) "Biological Rhythms and Psychiatry." In *American Handbook of Psychiatry,* Vol. 7, Chapter 3, pp. 46–74.

   D. Kripke (1984) "Critical Interval Hypothesis for Depression." *Chronobiology International,* 1(1): 73–81.

   ——— (1983) "Phase-Advance Theories for Affective Illnesses." In *Circadian Rhythms in Psychiatry,* edited by T. A. Wehr and F. K. Goodwin, pp. 41–69. Pacific Grove, Calif.: Boxwood Press.

5. E. L. Peterson (1980) "A Limit Cycle Interpretation of a Mosquito Circadian Oscillator." *Journal of Theoretical Biology,* 84: 281–310.

   E. L. Peterson (1980) "Phase-Resetting a Mosquito Circadian Oscillator." *Journal of Comparative Physiology,* 138: 201–211.

   E. L. Peterson (1981) "Dynamic Response of a Circadian Pacemaker," parts I and II. *Biological Cybernetics,* 40: 171–194.

The word *rhythm* used in this chapter and later refers to a measured quantity that varies roughly periodically in time. A rhythm is an observable manifestation of something happening with that same period inside an organism. That postulated "something" is called the circadian "clock." The overt rhythm may directly reveal some aspect of the clock process, or it may be only an indirect result of the clock process. The clock goes through a cycle of phases, repeating none. In contrast, a rhythmic quantity can only increase and decrease through the same values: it thus goes through every level at least twice in each cycle, once while increasing and once while decreasing. Underneath the plot of a rhythm we will sometimes lay a scale of phase, giving each phase of the cycle a unique color, even though the rhythm may pass through the same level at several distinct phases.

6. J. W. Hastings and B. Sweeney (1958) "A Persistent Diurnal Rhythm of Luminescence in *Gonyaulax polyedra.*" *Biological Bulletin,* 115: 440–458.

   Experiments at about the same time using the fruit fly *Drosophila pseudoobscura* were also producing data that I believe should have been construed as smooth even resetting:

   C. S. Pittendrigh and V. G. Bruce (1957) "An Oscillator Model for Biological Clocks." In *Rhythmic and Synthetic Processes in Growth,* edited by D. Rudnick, pp. 75–109. Princeton, N.J.: Princeton University Press.

Chapter Four

1. There are two ways we might define *hue* for present purposes. The one chosen here is the conventional definition of *hue* as one of the three independent attributes of any color, along with brilliance and saturation. Only the white-grey-black series of colors are hueless by this definition. The other definition is used for almost all arguments in *When Time Breaks Down* (Winfree 1987; see note 2 below): hue is a label on each color in the palette chosen to represent phases of any cycle, and "hueless" only means "not in the palette."

2. A. T. Winfree (1987) *When Time Breaks Down.* Princeton, N.J: Princeton University Press.

   S. H. Strogatz (1984) "Yeast Oscillations, Belousov-Zhabotinsky Waves, and the Non-retraction Theorem." *Mathematical Intelligencer,* 7(2): 9–17.

Chapter Five

1. M. S. Johnson (1939) "Effect of Continuous Light on Periodic Spontaneous Activity of White-Footed Mice (Peromyscus)." *Journal of Experimental Zoology,* 82: 315–328.

2. F. A. Brown, Jr., M. Fingerman, M. I. Sandeen, and H. M. Webb (1953) "Persistent Diurnal and Tidal Rhythms of Color Change in the Fiddler Crab, *Uca pugnax.*" *Journal of Experimental Zoology,* 123: 29–60.

   F. A. Brown, Jr., J. Shriner, and C. L. Ralph (1956) "Solar and Lunar Rhythmicity in the Rat in 'Constant Conditions' and the Mechanism of Physiological Time Measurement." *American Journal of Physiology,* 184: 491–496.

   C. S. Pittendrigh (1954) "On Temperature Independence in the Clock System Controlling Emergence Time in *Drosophila.*" *Proceedings of the National Academy of Sciences USA,* 40: 1018–1029.

   C. S. Pittendrigh (1960) "Circadian Rhythms and the Circadian Organization of Living Systems." In *Cold Spring Harbor Symposium on Quantitative Biology,* 25: 159–184.

3. K. S. Rawson (1956) "Homing Behavior and Endogenous Activity Rhythms." Ph.D. dissertation, Harvard University.

4. C. S. Pittendrigh and V. G. Bruce (1957) "An Oscillator Model for Biological Clocks." In *Rhythmic and Synthetic Processes in Growth,* edited by D. Rudnick, pp. 75–109. Princeton, N.J: Princeton University Press.

5. H.-S. Shin, T. A. Bargiello, B. T. Clark, F. R. Jackson, and M. W. Young (1985) "An Unusual Coding Sequence from a *Drosophila* Clock Gene Is Conserved in Vertebrates." *Nature,* 317: 445–448.

   F. R. Jackson, T. A. Bargiello, S.-H. Yun, and M. W. Young (1986) "Product of *per* Locus of *Drosophila* Shares Homology with Proteoglycans." *Nature,* 320: 185–188.

6. G. K. Chesterton attribution, "Mathematical Games," *Scientific American,* 236(1): 116.

7. E. L. Peterson and D. S. Saunders (1980) "The Circadian Eclosion Rhythm in *Sarcophaga argyostoma:* A Limit Cycle Representation of the Pacemaker." *Journal of Theoretical Biology,* 86: 265–277.

   E. L. Peterson (1980) "Phase-Resetting a Mosquito Circadian Oscillator." *Journal of Comparative Physiology,* 138: 201–211.

   E. L. Peterson (1981) "Dynamic Response of a Circadian Pacemaker." Parts I and II. *Biological Cybernetics,* 40: 171–194.

   A. T. Winfree and H. Gordon (1977) "The Photosensitivity of a Mutant Circadian Clock." *Journal of Comparative Physiology,* 122: 87–109.

8. W. Engelmann, I. Eger, A. Johnsson, and H. G. Karlsson (1974) "Effect of Temperature Pulses on the Petal Rhythm in *Kalanchoë:* An Experimental and Theoretical Study." *International Journal of Chronobiology,* 2: 347–358.

   W. Engelmann and A. Johnsson (1978) "Attenuation of the Petal Movement Rhythm in *Kalanchoë* with Light Pulses." *Physiologia Plantarum,* 43: 68–76.

9. W. Taylor, R. Krasnow, J. C. Dunlap, H. Broda, and J. W. Hastings (1982) "Critical Pulses of Anisomycin Drive the Circadian Oscillator in *Gonyaulax* Towards Its Singularity." *Journal of Comparative Physiology,* 148: 11–25.

   J. R. Malinowski, D. L. Laval-Martin, and L. N. Edmunds, Jr. (1985) "Circadian Oscillators, Cell Cycles, and Singularities: Light Perturbation of the Free-Running Rhythm of Cell Division in *Euglena.*" *Journal of Comparative Physiology,* B155: 257–276.

10. A. T. Winfree (1972) "Oscillatory Glycolysis in Yeast: The Pattern of Phase Resetting by Oxygen." *Archives of Biochemistry and Biophysics,* 149: 338–401.

    See also Chapter 7 of this book.

11. A. Johnson (1976) "Oscillatory Transpiration and Water Uptake in *Avena* Plants." *Bulletin of the Institute for Applied Mathematics,* 12: 22–26.

    A. Johnsson, T. Brogardh, and O. Holje (1979) "Oscillatory Transpiration of *Avena* Plants: Perturbation Experiments Provide Evidence for a Stable Point of Singularity." *Physiologia Plantarum,* 45: 393–398.

12. E. L. Peterson and L. Calabrese (1982) "Dynamic Analysis of a Rhythmic Neural Circuit in the Leech *Hirudo medicinalis.*" *Journal of Neurophysiology,* 47: 256–271.

13. J. Jalife and C. Antzelevitch (1979) "Phase Resetting and Annihilation of Pacemaker Activity in Cardiac Tissue." *Science,* 206: 696–697.

14. D. Paydarfar, F. L. Eldridge, and J. P. Kiley (1986) "Resetting of Mammalian Respiratory Rhythm: Existence of a Phase Singularity." *American Journal of Physiology,* 250, R721–R727.

    ——— and F. L. Eldridge (1986) "Phase Resetting and Dysrythmic Responses of the Respiratory Oscillator." *American Journal of Physiology,* in press.

15. A. T. Winfree (1977) "The Phase Control of Neural Pacemakers." *Science,* 197: 761–762.

16. A. T. Winfree (1977) *When Time Breaks Down.* Princeton, N.J: Princeton University Press.

    The figure is colored from the calculations of V. Reiner and C. Antzelevitch (1985) "Phase Resetting and Annihilation in a Mathematical Model of the Sinus Node," *American Journal of Physiology,* 249: H1143–H1153.

17. See note 16 above.

18. P.-S. Chen, N. Shibata, E. G. Dixon, P. Wolf, N. Daniely, M. Sweeney, W. Smith, and R. E. Ideker (1986) "Activation During Ventricular Defibrillation in Open Chest Dogs: Evidence of Complete Cessation and Regeneration of Ventricular Fibrillation after Unsuccessful Shocks." *Journal of Clinical Investigations,* 77: 810–823.

    P.-S. Chen, N. Shibata, E. G. Dixon, R. O. Martin, and R. E. Ideker (1986) "Comparison of the Defibrillation Threshold and the Upper Limit of Ventricular Vulnerability." *Circulation,* 73(5): 1022–1028.

N. Shibata, P.-S. Chen, E. Summers, P. Wolf, N. D. Daniely, J. D. Spoon, W. Smith, and R. E. Ideker (1985) "Epicardial Activation after Shocks in the Vulnerable Period and in Ventricular Fibrillation." *Circulation,* 72 (Supplement III): 958.

N. Shibata, P.-S. Chen, S. J. Worley, E. Summers, D. J. Stilwell, and R. E. Ideker (1985) "Shock Strength and Ventricular Vulnerability." *Circulation,* 72 (Supplement III): 1528.

N. Shibata, P.-S. Chen, and R. E. Ideker (1986) "The Initiation of Ventricular Fibrillation During the Vulnerable Period." *Clinical Progress in Electrophysiology and Pacing,* in press.

Chapter Six

1. For an understanding of the cuckoo-clock limit cycle, see:

   N. Minorsky (1962) *Nonlinear Oscillations.* Princeton, N.J: Van Nostrand.

   A. A. Andronov, A. A. Vitt, and S. E. Khaikin (1966) *Theory of Oscillators.* Oxford: Pergamon Press.

   R. Zwanzig (1976) "Interactions of Limit Cycle Oscillators." In *Topics in Statistical Mechanics and Biophysics,* edited by R. A. Piccirelli. American Institute of Physics.

2. A. T. Winfree (1980) *The Geometry of Biological Time.* New York: Springer-Verlag, Chapters 7 and 20.

3. A. T. Winfree and H. Gordon (1977) "The Photosensitivity of a Mutant Circadian Clock." *Journal of Comparative Physiology,* 122: 87–109.

4. E. L. Peterson (1980) "Phase-Resetting a Mosquito Circadian Oscillator." *Journal of Comparative Physiology,* 138: 201–211.

   E. L. Peterson (1981) "Dynamic Response of a Circadian Pacemaker." Parts I and II. *Biological Cybernetics,* 40: 171–194.

5. C. A. Czeisler (1978) "Human Circadian Physiology: Internal Organization of Temperature, Sleep-Wake, and Neuroendocrine Rhythms." Ph.D. dissertation, Stanford University.

   Alternations of sleep with waking at a period much longer than circadian was first reported by J. Aschoff in "Circadian Rhythms in Man." *Science,* 148: 1427–1432 (1965). R. Wever began investigations of substantially irregular alternations, punctuated by occasional extraordinary intervals of sleep or no sleep: "The Circadian Multioscillator System of Man." *International Journal of Chronobiology,* 3: 19–55 (1975).

6. W. Engelmann, H. G. Karlsson, and A. Johnsson (1973) "Phase Shifts in the *Kalanchoë* Petal Rhythm Caused by Light Pulses of Different Durations." *International Journal of Chronobiology,* 1: 147–156. Original data, courtesy of W. Engelmann, replotted as a three-dimensional time crystal.

7. J. R. Malinowski, D. L. Laval-Martin, and L. N. Edmunds, Jr. (1985) "Circadian Oscillators, Cell Cycles, and Singularities: Light Perturbation of the Free-Running Rhythm of Cell Division in *Euglena.*" *Journal of Comparative Physiology,* B155: 257–276.

8. L. McMurry and J. W. Hastings (1972) "No Desynchronization among Four Circadian Rhythms in the Unicellular Alga *Gonyaulax polyedra.*" *Science,* 175: 1137–1139.

9. The notion that cells might mutually synchronize their circadian clocks was first posed theoretically in a paper by C. S. Pittendrigh and V. G. Bruce (1957) "An Oscillator Model for Biological Clocks," in *Rhythmic and Synthetic Processes in Growth,* edited by D. Rudnick, pp. 75–109, Princeton, N.J: Princeton University Press. Failure met every attempt to experimentally observe such interactions, up to and including an experiment by F. M. Sulzman, V. D. Gooch, K. Homma, and J. W. Hastings (1982) "Cellular Autonomy of the *Gonyaulax* Circadian Clock," *Cell Biophysics,* 4: 97–103.

10. J. W. Hastings, H. Broda, and C. H. Johnson (1985) "Phase and Period Effects of Physical and Chemical Factors. Do Cells Communicate?" In *Temporal Order,* edited by L. Rensing and N. I. Jaeger, pp. 213–221. Berlin: Springer-Verlag.

11. D. Njus, V. D. Gooch, and J. W. Hastings (1981) "Precision of the *Gonyaulax* Circadian Clock." *Cell Biophysics,* 3: 223–231.

12. See note 11 above.

13. B. Walz and B. M. Sweeney (1979) "Kinetics of the Cyclohexamide-induced Phase Changes in the Biological Clock in *Gonyaulax.*" *Proceedings of the National Academy of Sciences USA,* 76: 6443–6447.

14. W. Taylor, R. Krasnow, J. C. Dunlap, H. Broda, and J. W. Hastings (1982) "Critical Pulses of Anisomycin Drive the Circadian Oscillator in *Gonyaulax* Towards Its Singularity." *Journal of Comparative Physiology,* 148: 11–25. The righthand figure on page 115 is a mathematical function contrived by the author to conform within two hours to these data; I

thank J. W. Hastings and R. Krasnow for access to the original data.

15. See note 14 above.

16. See note 2 above.

Chapter Seven

1. This was measured with care in 1972 by the same procedure as in D. Njus, V. D. Gooch, and J. W. Hastings (1981) "Precision of the *Gonyaulax* Circadian Clock," *Cell Biophysics*, 3: 223–231. The paper and all data were shredded by the U.S. Post Office, but the fact remains.

2. See references in pages 394–399 of A. T. Winfree (1980) *The Geometry of Biological Time.* New York: Springer-Verlag.

3. E. Bunning (1973) *The Physiological Clock.* New York: Springer-Verlag, p. 38.

4. A. T. Winfree (1972) "Oscillatory Glycolysis in Yeast: The Pattern of Phase Resetting by Oxygen." *Archives of Biochemistry and Biophysics*, 149: 338–401.

5. A. T. Winfree (1967) "Biological Rhythms and the Behavior of Populations of Coupled Oscillators." *Journal of Theoretical Biology*, 16: 15–42.

   ——— (1971) "Comment on Multioscillator Splitting." In *Biochronometry*, edited by M. Menaker, pp. 151–152. Washington, D.C: National Academy of Sciences.

6. See note 5 above and "The Firefly Machine," Chapter 11 of the source in note 2.

7. J. T. Enright (1980) *The Timing of Sleep and Wakefulness.* Berlin: Springer-Verlag.

   ——— and A. T. Winfree (1986) "Detection of a Singularity in a Coupled Stochastic System." In *Lectures on Mathematics in the Life Sciences*, edited by G. Carpenter. Providence, R.I.: American Mathematical Society.

8. S. Kauffman and J. J. Wille (1976) "The Mitotic Oscillator in *Physarum polycephalum.*" *Journal of Theoretical Biology*, 55: 47–93.

   J. J. Tyson and W. Sachsenmaier (1978) "Is the Nuclear Division in *Physarum* Controlled by a Continuous Limit-Cycle Oscillator?" *Journal of Theoretical Biology*, 73: 723–737.

9. A. K. Ghosh, B. Chance, and E. K. Pye (1971) "Metabolic Coupling and Synchronization of NADH Oscillations in Yeast Cell Populations." *Archives of Biochemistry and Biophysics*, 145: 319–331.

10. See note 9 above.

11. This figure and the one on page 139 are computed from a slightly more symmetric idealization of the phase compromise mechanism derived in A. T. Winfree (1974) "Patterns of Phase Compromise in Biological Cycles," *Journal of Mathematical Biology*, 1: 73–95. The figures on pages 125 and 134–135 are very similar to this idealization but are hand-drawn for closer quantitative fit to the data of the particular experiments described.

12. A. T. Winfree (1974) "Patterns of Phase Compromise in Biological Cycles." *Journal of Mathematical Biology*, 1: 73–95. The figures on pages 136–137 and 139 are based on the asymmetric model of Figure 10 in that paper in "recessive" framing of the cycle (phase zero taken as ¼ cycle after NADH maximum). The figure on page 140 is from the model, modified to better fit the data of the figures on pages 134 and 135 in "dominant" framing of the cycle (phase zero taken as ¼ cycle before NADH maximum).

13. See note 9 above.

14. Theodore H. Bullock (1955) "Compensation for Temperature in the Metabolism and Activity of Poikilotherms." *Biological Reviews of the Cambridge Philosophical Society*, 30: 311–342.

    See also other references on page 381 of A. T. Winfree (1980) *The Geometry of Biological Time*, New York: Springer-Verlag.

15. J. F. Feldman (1985) "Genetic and Physiological Analysis of a Circadian Clock Gene in *Neurospora crassa.*" In *Temporal Order*, edited by L. Rensing and N. I. Jaeger. New York: Springer-Verlag.

16. P. L. Lakin-Thomas and S. Brody (1985) "Circadian Rhythms in *Neurospora crassa:* Interactions Between Clock Mutations." *Genetics*, 109: 49–66.

    Cote and S. Brody (1986) "Circadian Rhythms in *Drosophila melanogaster:* Analysis of Period as a Function of Gene Dosage at the *per* (period) Locus." *Journal of Theoretical Biology*, in press.

17. H.-S. Shin, T. A. Bargiello, B. T. Clark, F. R. Jackson, and M. W. Young (1985) "An Unusual Coding Sequence from a *Drosophila* Clock Gene Is Conserved in Vertebrates." *Nature*, 317: 445–448.

    F. R. Jackson, T. A. Bargiello, S.-H. Yun, and M. W. Young (1986) "Product of *per* Locus of *Drosophila* Shares Homology with Proteoglycans." *Nature*, 320: 185–188.

18. S. Brody, C. Dieckmann, and S. Mikolajczyk (1985) "Circadian

Rhythms in *Neurospora crassa:* The Effects of Point Mutations on the Proteolipid Portion of the Mitochondrial ATP Synthetase." *Molecular and General Genetics,* 200: 155–161.

Chapter Eight

1. L. Glass and A. T. Winfree (1984) "Discontinuities in Phase-Resetting Experiments." *American Journal of Physiology,* 246: R251–R258.
2. H. Kalmus and L. A. Wigglesworth (1960) "Shock-excited Systems as Models for Biological Rhythms." In *Cold Spring Harbor Symposia on Quantitative Biology 25:* 211–216.

   R. Wever (1962–64) "Zum Mechanismus der biologischen 24-Stunden-Periodik." *Kybernetik,* 1: 139–154, 213–231, and 2: 127–144.
3. C. Pittendrigh (1967) "Circadian Rhythms, Space Research, and Manned Space Flight." In *Life Sciences and Space Research 5,* pp. 122–134. Amsterdam: North-Holland.
4. A. T. Winfree (1976) "The Morning Glory's Strange Behavior." *Horticulture,* 54(4): 42–51.

Chapter Nine

1. This chapter is essentially the abstract of a more technical monograph that develops the subject of phase singularities in spatial context: A. T. Winfree (January 1987) *When Time Breaks Down: The Three-dimensional Dynamics of Electrochemical Waves and Cardiac Arrhythmias.* Princeton, N.J: Princeton University Press. Citations can be found there: only a few are noted explicitly in Chapter 9 here.
2. D. Paydarfar, F. L. Eldridge, and J. P. Kiley (1986) "Resetting of Mammalian Respiratory Rhythm: Existence of a Phase Singularity." *American Journal of Physiology,* 250: R721–R727.

   D. Paydarfar and F. L. Eldridge (1986) "Phase Resetting and Dysrythmic Responses of the Respiratory Oscillator." *American Journal of Physiology,* in press.
3. H. R. Wilson and J. D. Cowan (1972) "Excitatory and Inhibitory Interactions in Localized Populations of Model Neurons." *Biophysical Journal,* 12: 1–24.
4. J. T. Enright (1980) *The Timing of Sleep and Wakefulness.* Berlin: Springer-Verlag.

   J. T. Enright and A. T. Winfree (1986) "Detection of a Singularity in a Coupled Stochastic System." In *Lectures on Mathematics in the Life Sciences,* edited by G. Carpenter. Providence, R.I.: American Mathematical Society.
5. A. T. Winfree and G. Twaddle (1981) "The *Neurospora* Mycelium as a Two-dimensional Sheet of Coupled Circadian Clocks." In *Mathematical Biology,* edited by T. A. Burton. New York: Pergamon Press.
6. A demonstration kit of dried membrane impregnated with this reagent can be ordered from Liquidfire, P.O. Box 43236, Tucson, AZ, 85733. As of this writing, the price is $10.00. For suggestions on how to use the kit, see "The Amateur Scientist," *Scientific American,* July 1978, pp. 152–158.
7. P.-S. Chen, N. Shibata, E. G. Dixon, P. Wolf, N. Daniely, M. Sweeney, W. Smith, and R. E. Ideker (1986) "Activation During Ventricular Defibrillation in Open Chest Dogs: Evidence of Complete Cessation and Regeneration of Ventricular Fibrillation after Unsuccessful Shocks." *Journal of Clinical Investigations,* 77: 810–823.

   P.-S. Chen, N. Shibata, E. G. Dixon, R. O. Martin, and R. E. Ideker (1986) "Comparison of the Defibrillation Threshold and the Upper Limit of Ventricular Vulnerability." *Circulation,* 73(5): 1022–1028.

   N. Shibata, P.-S. Chen, E. Summers, P. Wolf, N. D. Daniely, J. D. Spoon, W. Smith and R. E. Ideker (1985) "Epicardial Activation after Shocks in the Vulnerable Period and in Ventricular Fibrillation." *Circulation,* 72 (Supplement III): 958.

   N. Shibata, P.-S. Chen, S. J. Worley, E. Summers, D. J. Stilwell, and R. E. Ideker (1985) "Shock Strength and Ventricular Vulnerability." *Circulation,* 72 (Supplement III): 1528.

   N. Shibata, P.-S. Chen, and R. E. Ideker (1986) "The Initiation of Ventricular Fibrillation during the Vulnerable Period." *Clinical Progress in Electrophysiology and Pacing,* in press.
8. There is only one report of even resetting in any vertebrate circadian clock (a bird's): A. Eskin (1971) "Some Properties of the System Controlling the Circadian Activity Rhythm in Sparrows." In *Biochronometry,* edited by M. Menaker, pp. 55–80. Washington, D.C.: National Academy of Sciences. Unless vertebrate (including human) clocks operate on utterly different principles from all others, this shortage of examples will be corrected in the laboratory. A good starting place might be the human clock that regulates core temperature and waking time.

Sources of Illustrations

Drawings by Tom Cardamone Associates

Airbrushing by Ann Neumann

Facing page 1: Courtesy of the Beinecke Rare Book and Manuscript Library, Yale University.

Page 4: Courtesy of the National Air and Space Museum, Smithsonian Institution. Hand coloring by Travis Amos.

Page 14: from W. Whewell (1836) "Researches on the Tides." *Philosophical Transactions of the Royal Society of London,* 126: 289–307.

Page 16: Map from E. Schwiderski (1983) "Atlas of Ocean Tidal Charts and Maps. Part I: The Semidiurnal Principal Lunar Tide M_2." *Marine Geodesy,* 6(3–4): 219–265.

Page 20: Dennis de Cicco

Page 22: Arthur T. Winfree

Page 24: From R. E. Kronauer (1984) "Modeling Principles of Human Circadian Rhythms." In *Mathematical Modeling of Circadian Systems,* edited by M. C. Moore-Ede and C. A. Czeisler, pp. 105–128. New York: Raven Press.

Page 26: Adapted from J. Aschoff and R. Wever (1981) "The Circadian System of Man." In *Handbook of Behavioral Neurobiology 4,* pp. 311–329.

Pages 28 and 29: From C. P. Richter (1968) "Inherent 24-hour and Lunar Clocks of a Primate: The Squirrel Monkey." *Communications in Behavioral Biology,* 1: 305–332.

Page 30: From L. E. M. Miles, D. M. Raynal, and M. R. Wilson (1977) "Blind Man Living in Normal Society Has Circadian Rhythms of 24.9 Hours." *Science,* 198: 421–423. © 1977 by the AAAS.

Page 31: Adapted from N. Kleitman and T. G. Engelmann (1953) "Sleep Characteristics of Infants." *Journal of Applied Physiology,* 6: 269–282.

Page 32: Adapted from C. P. Kokkoris, E. D. Weitzman, C. P. Pollak, A. J. Spielman, C. A. Czeisler, and H. Bradlow (1978) "Long-term Ambulatory Temperature Monitoring in a Subject with a Hypernychthermal Sleep-Wake Cycle Disturbance. *Sleep,* 1: 177–190.

Page 34: Adapted from A. L. Weber, M. S. Cary, N. Connor, and P. Keyes (1980) "Human Non-24-Hour Sleep-Wake Cycles in an Everyday Environment." *Sleep,* 2:347–354.

Pages 35, 38, 39, and 40: From A. T. Winfree (1982) "Circadian Timing of Sleep and Wakefulness in Men and Women." *American Journal of Physiology,* 243: R193–R204.

Page 46: Charles Arenson

Page 48: Travis Amos

Page 49 (top): From S.-I. T. Inouye and H. Kawamura (1982) "Characteristics of a Circadian Pacemaker in the Suprachiasmatic Nucleus." *Journal of Comparative Physiology,* 146: 153–160.

Page 49 (bottom): From W. J. Schwartz, C. B. Smith, and L. C. Davidsen (1979) "In Vivo Glucose Utilization of the Suprachiasmatic Nucleus." In *Biological Rhythms and their Central Mechanism,* edited by M. Suda, O. Hayaishi, and H. Nakasawa, pp. 355–367. Amsterdam: Elsevier/North-Holland.

Page 50: Chip Clark

Pages 51 and 54: Redrawn from E. L. Peterson (1980) "A Limit Cycle Interpretation of a Mosquito Circadian Oscillator." *Journal of Theoretical Biology,* 84: 281–310.

Page 58: Adapted from A. T. Winfree (1973) "Resetting the Amplitude of *Drosophila's* Circadian Chronometer." *Journal of Comparative Physiology,* 85: 105–160.

Page 64: Adapted from J. W. Hastings and B. M. Sweeney (1958) "A Persistent Diurnal Rhythm of Luminescence in *Gonyaulax polyedra.*" *Biological Bulletin,* 115: 440–458.

Page 66: Courtesy of Beatrice Sweeney.

Page 70: Redrawn from E. L. Peterson (1980) "A Limit Cycle Interpretation of a Mosquito Circadian Oscillator." *Journal of Theoretical Biology,* 84: 281–310.

Pages 80, 81, and 82: Prepared by Arthur T. Winfree at Los Alamos National Laboratory with the support of the Department of Energy.

Page 84 (left): S. Stammers/Science Photo Library/Photo Researchers

Page 84 (right): Darwin Dale

Page 88: Arthur T. Winfree

Pages 90 (right) and 92: From A. T. Winfree (1980) *The Geometry of Biological Time.* New York: Springer-Verlag.

Page 94: Prepared by Arthur T. Winfree at Los Alamos National Laboratory with the support of the Department of Energy.

Page 95: Travis Amos

Page 97: Colored from contours computed in V. Reiner and C. Atnzelevitch (1985) "Phase Resetting and Annihilation in a Mathematical Model of the Sinus Node." *American Journal of Physiology,* 249: H1143–H1153.

Page 98: Wolfgang Engelmann

Page 101: Russ Kinne/Photo Researchers

Page 102: Adapted from A. T. Winfree (1970) "The Temporal Morphology of a Biological Clock." In *Lectures on Mathematics in the Life Sciences 2,* edited by M. Gerstenhaber, pp. 109–150. Providence, R.I.: American Mathematical Society.

Page 104: From E. L. Peterson (1981) "Dynamic Response of a Circadian Pacemaker, Part II." *Biological Cybernetics,* 40: 181–194.

Page 108: Wolfgang Engelmann

Page 109 (left): Adapted from A. T. Winfree (1980) *The Geometry of Biological Time.* New York: Springer-Verlag, p. 59.

Page 109 (right): W. Engelmann and A. Johnsson (1978) "Attenuation of the Petal Movement Rhythm in *Kalanchoë* with Light Pulses." *Physiologia Plantarum,* 43: 73.

Page 111: Eric Grave/Science Source/Photo Researchers

Page 115 (left): Adapted from page 19 of Taylor, R. Krasnow, J. C. Dunlap, H. Broda, and J. W. Hastings (1982) "Critical Pulses of Anisomycin Drive the Circadian Oscillator in *Gonyaulax* Towards Its Singularity." *Journal of Comparative Physiology,* 148: 11–25.

Page 115 (right): Computed from the data of Taylor et al. (see above). Data kindly provided by J. W. Hastings. This is a computer-assisted update of their page 18 contour map.

Page 118: John Shaw

Page 121: From E. K. Pye (1971) "Periodicities in Intermediary Metabolism." In *Biochronometry,* edited by M. Menaker, p. 627. Washington, D.C.: National Academy of Sciences.

Page 122: From A. Betz and B. Chance (1965) "Phase Relationship of Glycolytic Intermediates in Yeast Cells with Oscillatory Metabolic Control." *Archives of Biochemistry and Biophysics,* 109: 585–594.

Page 123: From A. T. Winfree (1980) *The Geometry of Biological Time.* New York: Springer-Verlag, p. 61.

Page 128: Arthur T. Winfree

Page 132: Patrick W. Grace/Photo Researchers

Pages 136, 137, and 139: Prepared by Arthur T. Winfree at Los Alamos National Laboratory with the support of the Department of Energy.

Page 144: NASA

Page 166: Fritz Goro

Page 168 (top): From D. Paydarfar, F. L. Eldridge, and J. P. Kiley (1986) "Resetting of Mammalian Respiratory Rhythm: Existence of a Phase Singularity." *American Journal of Physiology,* 250: R721–R727.

Page 168 (bottom): Colored following contours drawn in D. Paydarfar et al. (see above).

Page 172 (left and center): Reprinted with permission of the present publisher, Jones and Bartlett Publishers, Inc., from Shih and Kessel: *Living Images,* Science Books International, 1982, p. 23.

Pages 172 (right) and 173: Arthur T. Winfree

Page 174: Photographed by Fritz Goro in the laboratory of the author.

Page 175 (top left): Courtesy of Peter Newell

Page 175 (top right): Arthur T. Winfree

Page 175 (bottom left and right): Courtesy of Kenneth Raper

Page 176 (left): From B. Welsh (1984) "Pattern Formation in the Belousov-Zhabotinsky Reaction." Ph.D. dissertation, Glasgow College of Technology.

Page 177 (right): Prepared by Arthur T. Winfree at Los Alamos National Laboratory with the support of the Department of Energy.

Page 177: Adapted from A. T. Winfree (1983) "Sudden Cardiac Death: A Problem in Topology?" *Scientific American,* 248(5): 144–161. © *Scientific American,* 1983. All rights reserved.

Index

Other Books in the Scientific American Library Series

POWERS OF TEN
by Philip and Phylis Morrison and
the Office of Charles and Ray Eames

HUMAN DIVERSITY
by Richard Lewontin

THE DISCOVERY OF SUBATOMIC PARTICLES
by Steven Weinberg

THE SCIENCE OF MUSICAL SOUND
by John R. Pierce

FOSSILS AND THE HISTORY OF LIFE
by George Gaylord Simpson

THE SOLAR SYSTEM
by Roman Smoluchowski

ON SIZE AND LIFE
by Thomas A. McMahon and John Tyler Bonner

PERCEPTION
by Irvin Rock

CONSTRUCTING THE UNIVERSE
by David Layzer

THE SECOND LAW
by P. W. Atkins

THE LIVING CELL, VOLUMES I AND II
by Christian de Duve

MATHEMATICS AND OPTIMAL FORM
by Stefan Hildebrandt and Anthony Tromba

FIRE
by John W. Lyons

SUN AND EARTH
by Herbert Friedman

EINSTEIN'S LEGACY
by Julian Schwinger

ISLANDS
by H. W. Menard

DRUGS AND THE BRAIN
by Solomon H. Synder